Lean and Agile Project Management

T0271462

Lean and Agile Project Management

How to Make Any Project Better, Faster, and More Cost Effective
Second Edition

Terra Vanzant Stern, PhD

A PRODUCTIVITY PRESS BOOK

First published 2020
by Routledge
52 Vanderbilt Avenue, New York, NY 10017

and by Routledge
2 Park Square, Milton Park, Abingdon, Oxon, OX14 4RN

Routledge is an imprint of the Taylor & Francis Group, an informa business

Library of Congress Cataloging-in-Publication Data

Names: Vanzant Stern, Terra, author.
Title: Lean and agile project management : how to make any project better, faster, and more cost effective, second edition / Terra Vanzant Stern, PhD.
Description: Second Edition. | New York : Routledge, 2020. | Revised edition of the author's Lean and agile project management, [2017] | Includes bibliographical references and index.
Identifiers: LCCN 2020000030 (print) | LCCN 2020000031 (ebook) | ISBN 9780367359584 (hardback) | ISBN 9780429343414 (ebook)
Subjects: LCSH: Agile project management. | Project management. | Lean manufacturing. | Cost effectiveness. | Six sigma (Quality control standard)
Classification: LCC HD69.P75 V386 2020 (print) | LCC HD69.P75 (ebook) | DDC 658.4/013--dc23
LC record available at https://lccn.loc.gov/2020000030
LC ebook record available at https://lccn.loc.gov/2020000031

ISBN: 978-0-367-35958-4 (hbk)
ISBN: 978-0-429-34341-4 (ebk)

Typeset in Garamond
by Deanta Global Publishing, Services, Chennai, India

This book is dedicated to the Project Management Institute Mile-Hi Chapter (PMI Mile Hi) and to the Lean Enterprise Division of ASQ as well as the ASQ Denver Chapter. A special thank you to my clients, students and professional associations who support blending methodologies to create hybrid systems of thinking.

Contents

Foreword

"Don't say it cannot be done, rather say, you don't know how to do it yet."

Tomáš Baťa

With recent evolutions in Agile/Scaled Agile/Disciplined Agile, there have been so many myths created in project management communities. One is that traditional project management is fading away and all the latest projects are being executed using Agile frameworks or Scaled Agile principles only. As stated, this is only a myth. Recent evolutions have not clearly addressed the basic execution process and not ensured the coverage of all traditional processes.

This is one of the reasons that Dr. Stern's book is so important. It clearly spells out the individual methodologies and frameworks that impact project management. This gives the project manager the ability to create customized methodologies without losing credibility.

Here is an example of how I blended methodologies to create a hybrid project management system which I have named Rapid Fall Execution (RPE). Let's consider the exercise below:

> A boss walks to an employee and says, "Hi Susan, I need to you to fly to our headquarters in Atlanta office today and give a presentation to sales leadership tomorrow."

In this use case, Susan got less than 24 hours to plan and execute the task given by her boss. Does it resonate with any task/user story in Agile frameworks or Kanban boards? I believe, yes.

Let us take a look at what Susan does from the moment her boss leaves to the moment she completed presenting to her sales leadership or even to when she comes back from the trip.

Check if this trip is feasible to make in terms of family, health, weather, etc.

Call travel agent to secure the air tickets.

Reserve a hotel.

Arrange cabs for commuting to airports and offices.

Pack bags.

Gather all office material required for the presentations.

Check in and print boarding pass/keep it on the phone.

Take care of family arrangements for the duration of the trip.

Even though the trip is executed at a faster pace or in an Agile fashion, none of the steps are being missed or ignored. Let us try to map them to our traditional processes/knowledge areas.

Initiation—starting the trip arrangements.

Planning—booking the flights, cabs, hotel, and packing bags/office material.

Execution—giving the presentation to the salespeople and completing the trip.

Monitoring and controlling—making sure the presentations and trip are going as expected.

Closing—submitting the necessary reports, debriefing the boss, submitting the financials of the trip.

Scope management—what to be done in this trip? Planning for presentations and networking, etc.

Schedule management—going on time to catch flights, going to office, giving the presentation in the scheduled slot, and coming back home by catching return flight booked.

Cost management—expenditure of the trip is managed.

Quality management—providing value to the customers in this trip.

Procurement management—purchasing the essentials required for the trip.

Human resource management—Talking to the people and managing the people to make this trip valuable.

Communications management—90% of the job of any project manager is communication. Communicating with all the parties involved in making this trip successful.

In the above exercise or example, you can clearly see that nothing is skipped from traditional project management practices. Speedy execution of any project does not skip the steps. Granted, many Agile or Lean Agile practices eliminate redundancy or waste, but they do not skip out the essentials.

In any Agile framework, the grassroot execution process is not very clear. Companies and teams should find a pragmatic way to implement projects as per their culture. As we all know, being Agile is more important than doing Agile. Many Agile practitioners and companies provide frameworks which can be brought in as a whole into the execution model but can/have to be tweaked according to how the company or department works. There is no one-size-fits-all Agile framework or practice to give a guaranteed delivery/success for the company.

As Peter Drucker says, "culture eats strategy for breakfast." Do not create strategies that tactically operate in a different direction to culture.

Current market trends in project management and the offerings outside give many ideas to take in that may lead to a better way of executing any project in companies.

Rapid Fall Execution is a way I personally was able to implement a Waterfall project using Agile techniques. RPE allows for closer collaboration with the customer. It capitalizes on iterative and dynamic functionality.

This book provides a foundation which will allow you to create customized methodologies as well.

Krishna Anantharaju, PMP, CSM, SA, POPM, SSBBP, ITIL v3
President and CEO, PMI Mile Hi Chapter

Preface

"I don't see how he can ever finish, if he doesn't begin."

**Lewis Carroll, *Alice's Adventures in Wonderland*
and *Through the Looking-Glass***

We have learned a lot since the first edition of *Lean and Agile Project Management* was published in 2017. This is primarily because more businesses are starting to embrace better ways of doing things. We no longer fear the idea of blending several business concepts to achieve a hybrid methodology that might better meet the needs of a specific industry. In my original work, I opened with "Imagination is the only weapon in the war against reality," a quote I adore from Lewis Carroll's work, *Alice's Adventures in Wonderland*. When I first started the second edition of this book, I revisited the quote. What did Alice's journey teach us about process improvement? Why is imagination the first step to making things better, faster, or more cost-effective? The answer is actually very simple. I think Alice, herself, would approve: To think outside the box, first you must learn how to open the box.

Alice, the seven-year-old protagonist in *Alice's Adventures in Wonderland*, and I have a lot in common. We both have an unquenchable curiosity about our surroundings. We are both abnormally interested in hearing the stories of others. And we both began our life journeys believing the world was an orderly and stable place.

I grew up in a military family and joined the military myself shortly after high school. As a child and young woman, I was primarily exposed to a structured and predictable world. Before Alice fell down the rabbit hole, she was living in the confines of a well-run estate. The Wonderland-like scenarios I later encountered in the civilian corporate workforce fascinated and frustrated me almost as much as Wonderland itself did Alice. The world relies on a certain amount of steadiness to run properly. As a project manager, I often had to reason with Red Queens and Mad Hatters to accomplish my project goals.

Our newest generation does not always recognize the customary way of doing things as the best approach. They sometimes require a certain amount of foolishness and lack of structure to stay engaged. This is counterintuitive to mature project managers interested in staying in the game. We have to continually convince our teams that anyone who truly wants to be innovative must first understand the rules and customs of their antecedents. It takes both conventional wisdom and nonconventional philosophy to finish a project on time and on budget.

Applying Lean concepts and Agile techniques to traditional project management is the best way to capitalize on our creativity and still respect the core project management science that has served us well since it emerged as a respected discipline in the 1950s. Lean and Agile provide an avenue for multiple generations to work together.

Alice teaches us some crucial lessons that can be applied today as we move through managing projects. For example, one famous revelation on her journey is that she becomes aware that "It's no use going back to yesterday, because I was a different person then." Instead, we need to grow and change as we acquire new information. Traditional PMs sometimes rely too much on what worked well yesterday. New knowledge needs to be explored and often incorporated.

Alice also discovered that each of us has a story to tell from the lens and perspective of our own experiences, and that is an important part of self-fulfillment: "There ought to be a book written about me, that there ought. And when I grow up, I'll write one." I may be stretching Lewis Carroll's literary intentions when he wrote the book in 1865, but I think this means we need to appreciate the stories of others. Everyone owns a brilliant piece of the puzzle that just might fit. Traditional project management is often limited by the "my thoughts and plans are the only ones that matter" mentality.

Alice generally gave herself solid advice but seldom followed it, which simply means we must listen from both our hearts and heads when we are involved in problem solving or project management. We generally know what is right and in the best interest of the project. This is one of the reasons why the project plan is so important. The plan keeps us on track and helps us remember our short- and long-term goals. It reminds us of the deliverables and helps us celebrate our milestones.

But it turns out that people management is the most time-consuming function of all projects. There is no one-size-fits-all plan for that. However, Lean thinking and Agile techniques (Lean and Agile), when embraced by traditional project management, provide the opportunity to work within a solid framework, allowing the freedom to apply creativity on demand. But more importantly, they provide better vehicles to manage people.

In the past, a PM has had the luxury to depend on the human resources (HR) department for the people issues. Now, with smaller staffs, HR often is only visible at the beginning and at the end of the project. HR issues that arise during the course of a given project are handled on a first-come, first-served basis or put on hold. HR functions have become more comprehensive and complex. The PM has to know when and if the problem is serious enough to involve HR and how much time this inclusion could cost the project.

When a decision must be made immediately, a PM is required to switch roles and become the HR representative. Globalization has made management involvement more necessary and less transparent to the employee. It is no longer uncommon for a staff member to go directly to his or her PM when seeking HR advice.

The science of HR management covers a wide range. To suggest that the PM must learn everything there is to know is unreasonable. However, a PM who is familiar with HR issues will avoid a great deal of grief. Most HR issues focus on communication. Traditional project management does not embrace the issue of communication and change management as much as Lean and Agile.

The PM who is assigned to international projects must be aware of how business is done in different countries. This is another area in which Lean and Agile appear to be more malleable. The use of visual controls, flowcharts, and roadmaps is encouraged.

A PM needs to be able to interpret special nuances in the company's HR policy. A company may electively revise policies to protect areas, such as sexual orientation and religious rights, without realizing the two policies clash. In theory, the policy may seem correct, but it is the PM who is able to observe the real-life consequences. A PM can easily overstep his or her boundaries and be totally unaware. Likewise, the PM may be unwittingly caught in the middle of an employer-sponsored drive to involve employees in the political process or union activities.

A PM's communication skills now need to include more focus on the following:

Coaching
Mentoring
Mediation/arbitration
Dispute resolution

Additionally, a PM has to deal with certain aspects of employee development that once were handled by the company's training department. A PM is now expected to help his or her subordinates identify learning opportunities as well as formulate plans designed to realize and document expectations.

Cultural management and diversity issues are quickly becoming topics that a PM cannot ignore. The workforce has become older, causing generational differences. The workforce has become more international, creating the need to understand diversity. There is an increased emphasis on safety and security. It is a wonder that a PM has time to actually do a project with all the additional responsibilities.

Many times, as a PM, it is important to maintain imagination and inspiration when following a standard rollout plan even though there really isn't a lot of flexibility once a project plan is in place and a time/cost baseline has been created.

A PM can be subject to explaining why a particular project plan is becoming so hard to execute. Didn't everyone attend the same meeting? Did they not understand the task they were given or that there was a timeline involved? Why push back now when all the steps are in place? Just like Alice, a project manager can suddenly feel that the world has just turned upside down!

In addition to Alice being a curious soul, she was also clumsy and easily frustrated. As a PM, it is hard not to identify with these characteristics at least some of the time. The good news is that the Lean and Agile toolkit can replace what are normally exasperating circumstances with opportunities to make a project better. By applying or considering Lean and Agile as an option, day-to-day project activities can begin to feel less awkward and more graceful.

The Cheshire Cat told Alice that everyone in Wonderland was mad. This included Alice herself. However, as it turns out, a certain amount of madness or creativity is necessary today to stay fruitful. Alice's response to the Mad Hatter was, "You're entirely bonkers. But, I'll tell you a secret. All the best people are."

Terra Vanzant Stern, PhD, PMP, SPHR/GPHR
Six Sigma Master Black Belt and CEO
***Simple, Smart D**ecision-Making, Inc.*
www.SSDGlobal.net

About the Author

**Terra Vanzant Stern, PhD, PMP, SPHR/GPHR** is a Six Sigma Master Black Belt and technical writer. Her publications include books, white papers, and articles in the areas of leadership development, ethics, critical thinking, project management, and Lean Six Sigma.

Dr. Vanzant Stern is the CEO of Simple, Smart Decision-Making, Inc. dba, SSD Global Solutions (SSD) and the inventor of *Leaner* Six Sigma™ (L*r*SS)™. L*r*SS™ is a methodology used to simplify the popular process improvement theory, Lean Six Sigma. SSD is a federal government contractor who is also a preferred vendor of the State of Colorado. SSD's client list includes recognized· names such as Southern California Edison, Tennessee Valley Authority, Fidelity Insurance, and Blue Cross/Blue Shield. SSD has an international presence working in cities such as Douala, Cameroon and San Sebastian, Mexico.

Dr. Vanzant Stern served as the Chair of the ASQ Lean Enterprise Division as well as ASQ Denver Section. Prior board positions have also included Director, Strategic Planning, for the Colorado Human Resource Association (CHRA) and the Ethics Committee for the State of Colorado Office of Economic Development. She currently is on the board of the Colorado Lean Network. Presently, Dr. Stern is working on *The Business of Being Better: How Personal Integrity Can Shape the Future of Leadership and Project Management.*

Chapter 1

The Three Faces of Traditional Project Management

In order to truly understand how Lean concepts and Agile techniques (Lean and Agile) can be applied to project management, it is important to understand the accepted basics of project management. Most PMs are surprised to learn that there are three primary recognized bodies of knowledge for project management: PMBOK®, PRINCE2®, and ISO 21500. There are also a number of independent bodies that emphasize a particular area, such as health care or construction.

This work capitalizes on the parallels of the three primary bodies of knowledge and speaks to their similarities rather than their differences. However, a PM wishing to engage fully in Lean and Agile should know which of the three sources is being relied on in his or her organizations. Once this body of knowledge is established or adopted by the PM, the PM should become expertly familiar with that source.

The Project Management Body of Knowledge (PMBOK) is a US-based program supported by the Project Management Institute (PMI). It provides a set of standard terminology and guidelines. Although it overlaps with practices used in general management, there are a number of unique thoughts, such as critical path and work breakdown structure, not typically discussed in other management disciplines, such as financial forecasting or organizational development. PMI offers individual certification programs.

Projects in Controlled Environments, version 2 (PRINCE2) is a program that began as a joint venture between the UK government and a private company, Capita. PRINCE2 focuses on dividing projects into manageable and controllable phases. It encompasses quality management and also offers individual certification programs.

ISO 21500:2012, Guidance on Project Management is an international standard developed by the International Organization for Standardization (ISO). An interesting piece of trivia is that work began on this standard in 2007, but it was not officially published until 2012, making it the newest of the three primary bodies. This standard is the first of the intended series of standards, and it aligns by design with other more established standards in areas such as quality management systems (QMS) and risk management. Currently, it is considered a guideline, so there is no official certification or registration process.

One commonality in all three primary bodies is the concept of a project life cycle (PLC). The PLC refers to a series of activities that are necessary to fulfill project goals or objectives. It is more commonly known as the project management life cycle (PMLC).

The PMLC has five phases: initiation, planning, execution, monitoring, controlling, and closure. Some bodies of knowledge (BOK) do not specifically name these steps and/or combine them with other phases. Named or not, a successful project must go through each phase of the cycle.

To summarize the steps or phases in the cycle, the term initiation refers to selecting the goal. Planning the project involves estimating resources and time, identifying the order of tasks, determining the execution schedule, and performing a risk assessment. Execution, simply stated, involves performing the tasks. Monitoring and controlling occur during all phases of the project. They include monitoring resources, quality, risks, and overall project status. Closing is the phase that includes all the activities necessary for the project office to bring closure to the project effort.

Lean and Agile project management begins with closely observing each phase of the PMLC and envisioning Lean and Agile opportunities. This is a summary of the phases and how certain Lean and/or Agile tools might be used. A more in-depth discussion of each of the phases occurs later in this book.

Project Initiation

The project initiation phase is the most crucial phase in the PMLC. This phase establishes the scope. A major outcome is the project charter. A charter is typically developed by creating a business case followed by conducting a feasibility study. If there is more than one resource available to execute the project, a project team is established. There may also be a need to establish or partner with the project management office (PMO). There are several Lean opportunities.

A supplier-input-process-output-customer (SIPOC) analysis could be used in either the business case or feasibility study. This would identify all the stakeholders in the project and consider the nonhuman resources that may contribute to the success of the project. Using a template to create the project charter is another simple way to make the process Lean. The plan-do-check-act (PDCA) methodology may be the best way to establish a PMO.

Project Planning

Many aspects of project management come down to good planning. In the planning process, Lean and Agile do support the use or the awareness of management tools promoted first by total quality management (TQM) literature. These diagrams discussed later in this work include, but are not limited to, the following:

The KJ method or affinity diagrams
Interrelationship digraph (ID)
Tree diagrams
Prioritization diagrams
Matrix diagrams
Process decision diagrams
Activity network diagrams

There are a number of Lean and Agile tools that may be applied in this phase. For example, suggesting a 5S model, a five-step method of organizing and maintaining a workplace, prior to beginning the project may help in the execution if the environment is physically disorganized. Gemba walks, a term used to describe personal observations of work, can promote a greater understanding of constraints within the work environment.

A key performance indicator (KPI) is a business metric used to evaluate factors that are crucial to the success of an organization and can vary from organization to organization. Strongly promoted KPIs can be extremely powerful drivers of behavior and may be addressed in this phase of the cycle.

Many Agile professionals suggest five levels of planning to include the following:

Product or service vision
Product or service roadmap
Release or rollout plan
Sprint plan
A strategy to achieve daily commitment

Project Execution

Initiation and planning are necessary for efficacious execution of any project. Generally speaking, basic project management and Lean are in alignment with how a project should be deployed. The use of a project plan using a work breakdown structure (WBS) and establishing metrics are good examples. What Lean offers that is not typically addressed in project management methodologies is the mistake-proofing aspect. The strategy used to ensure the success of the project often involves placing controls and detection measures within the project plan.

Visual feedback systems (Andon) may encourage quicker execution. Creating continuous flow eliminates waste and speeds the process in many projects. Andon means sign or signal. It is a visual aid that alerts and highlights places where action is required; for example, a flashing light in a manufacturing plant that indicates the line has been stopped by one of the operators due to some irregularity.

In this phase, hoshin kanri, a policy deployment tool, may help ensure that progress toward the strategic goals is consistent. Hoshin kanri is a method for ensuring that the strategic goals of a company drive progress and action at every level within that company.

Project Monitoring and Controlling

The monitoring and controlling process oversees all the tasks and metrics necessary to ensure that the approved and authorized project is within scope, on time, and on budget so that the project proceeds with minimal risk. Lean and Agile promote more people interaction than typical project management. Project management models often rely more heavily on Gantt and other charts to track progress. Concentrating a little more on the people aspects will increase team accountability.

Lean specifically looks at poka-yoke and heijunka. Poka-yoke is the Japanese term for mistake proofing. Mistake proofing involves eliminating possibilities for errors. An example would be color-coding a wiring template to assist the worker. The input and output would be the same color.

Heijunka is the Japanese word for level scheduling. A level-scheduling strategy's objective is to minimize disruptions caused by sudden changes in demand levels by matching the product family schedules with product-by-product schedules. To achieve the objectives of level scheduling, both the sales and production departments must agree on a fixed level of output volume and output duration.

Project Closing

The purpose of the closing phase in the PMLC is to confirm completion of project deliverables to the satisfaction of the project sponsor and to communicate final project disposition and status to all participants and stakeholders. The concept of standardized work is often useful during this phase. This is when documented procedures capture best practices. If standardized work has been created, it may be used to accelerate the closing process.

Agile project closure is much more robust and has definitive objectives, such as handing the project over to operations, tidying up any loose ends, reviewing the project to a stronger extent, and making celebrating an essential activity as opposed to something that is nice to do.

Lean project closure concentrates on keeping improvements ongoing, documenting best practices, and encouraging "lessons learned" meetings. It embraces brainstorming around "how can we do it better next time."

The advantage of applying the Lean and Agile concepts to project management is that they incorporate stronger planning tools and various aspects of mistake proofing not classically addressed in basic project management theory. The journey begins with examining the PMLC through a Lean perspective. Lean project management focuses on making projects better, faster, and more cost-effective by eliminating waste and unnecessary activities.

The PMBOK specifically lists ten knowledge areas in project management that are addressed in a different structure within in PRINCE2 and ISO 21500:

1. Integration
2. Scope
3. Time management
4. Cost management

5. Quality
6. Human resources
7. Communications
8. Risk
9. Procurement
10. Stakeholders

These ten factors, as identified in the PMBOK, demonstrate how adding Lean and Agile to project management is beneficial.

PRINCE2, Change Management

One area that is more structured in PRINCE2 than other traditional project management methodologies is in change management. This excerpt from the PRINCE2 manual is a good summary:

> Changes to specification or scope can potentially ruin any project unless they are carefully controlled. Change is, however, highly likely. The control of change means the assessment of the impact of potential changes, their importance, their cost and a judgmental decision by management on whether to include them or not. Any approved changes must be reflected in any necessary corresponding change to schedule and budget. "Management" in the PRINCE2 context means either the Project Board or a subordinate group called the Change Authority. For simplicity I'll refer to the Change Authority in the description of the process except where Project Board is explicitly intended.

A chapter on the core concepts of project management would not be complete without some basic history.

In 2570 BCE, the Great Pyramid of Giza was completed. Although archaeologists still debate about how this amazing feat was achieved, most agree it is the first evidence of project management. There was some degree of planning, execution, and control involved in managing this project.

Fast-forwarding to 208 BCE sees the construction of the Great Wall of China. Although it is considered a wonder of the world, there is more documentation available about this project. Labor was organized into three groups, which included soldiers, common people, and criminals. In the end, Emperor Qin Shi Huang ordered millions of people to finish this project.

It was not until 1917 that people realized the use of a standard set of tools might be useful in project management. It started with the Gantt chart, developed by Henry Gantt. This tool was considered a major innovation in the 1920s. It was then used with much anticipation on the Hoover Dam project, which was started in 1931. With computerization and now much easier to use, Gantt charts are still in use today.

After the Hoover Dam project, project management as a science developed rapidly via these key events:

1956: The American Association of Cost Engineers (now AACE International).
1957: The critical path method (CPM) is invented by the DuPont Corporation.
1958: The program evaluation review technique (PERT) is invented for the US Navy's Polaris Project.
1962: US Department of Defense mandates the work breakdown structure (WBS) approach.
1965: The International Project Management Association (IPMA) is founded.
1969: Project Management Institute (PMI) is launched to promote the project management profession.

As project management matured, so did progressive thoughts and ideas. In 1984, Dr. Eliyahu Goldratt introduced a novel called *The Goal*. In this work, Dr. Goldratt addressed the theory of constraints (TOC). His premise was that any manageable system is limited in achieving more of its goal by a small number of constraints, and there is always at least one constraint. The TOC process is used to identify the constraint and examine how to exploit the obstacle. The methods and algorithms from TOC went on to form the basis of critical chain project management.

In 1986, Scrum became a recognized project management style, and in 1987, PMI released the first edition of the PMBOK. Three years later, in 1989, the PRINCE2 method was developed.

Roles and Responsibilities

Without exception, traditional project management BOKs agree about the roles and responsibilities of the PM and those who participate in the PMLC. This is a common summary considered standard thought in the industry.

Project Team

The project team is the group responsible for planning and executing the project. It consists of a PM and a variable number of project team members, who are brought in to deliver their tasks according to the project schedule.

The project team members are responsible for executing tasks and producing deliverables as outlined in the project plan and directed by the PM at whatever level of effort or participation has been defined for them.

On larger projects, some project team members may serve as team leads, providing task and technical leadership, and sometimes maintaining a portion of the project plan.

Project Manager

The PM is the person responsible for ensuring that the project team completes the project. The PM develops the project plan with the team and manages the team's performance of project tasks. It is also the responsibility of the PM to secure acceptance and approval of deliverables from the project sponsor and stakeholders. The PM is responsible for communication, including status reporting, risk management, escalation of issues that cannot be resolved by the team, and in general making sure the project is delivered on budget, on schedule, and within scope.

Sponsor

The sponsor is a manager with demonstrable interest in the outcome of the project, and he or she is ultimately responsible for securing spending authority and resources for the project. Ideally, the executive sponsor should be the highest-ranking manager possible in proportion to the project size and scope. The executive sponsor acts as a vocal and visible champion, legitimizes the project's goals and objectives, keeps abreast of major project activities, and is the ultimate decision maker for the project. The executive sponsor provides support for the project sponsor and/or project director and the PM, has final approval of all scope changes, and signs off on approvals to proceed to each succeeding project phase. The executive sponsor may elect to delegate some of the above responsibilities to the project sponsor and/or project director.

The project sponsor and/or project director is a manager with demonstrable interest in the outcome of the project, and he or she is responsible for securing spending authority and resources for the project. The project sponsor acts as a vocal and visible champion, legitimizes the project's goals and objectives, keeps abreast of major project activities, and is a decision maker for the project. The project sponsor will participate in and/or lead project initiation, the development of the project charter. He or she will participate in project planning (high level) and the development of the project initiation plan. The project sponsor provides support for the PM; assists with major issues, problems, and policy conflicts; removes obstacles; is active in planning the scope; approves scope changes; signs off on major deliverables; and signs off on approvals to proceed to each succeeding project phase. The project sponsor generally chairs the steering committee on large projects. The project sponsor may elect to delegate any of the above responsibilities to other personnel either on or outside the project team.

Steering Committee

The steering committee generally includes management representatives from the key organizations involved in the project oversight and control and any other key stakeholder groups that have special interest in the outcome of the project. The steering committee acts individually and collectively as a vocal and visible project champion throughout their representative organizations; generally they approve project deliverables, help resolve issues and policy decisions, approve scope changes, and provide direction and guidance to the project. Depending on how the project is organized, the steering committee can be involved in providing resources, assist in securing funding, act as liaisons to executive groups and sponsors, and fill other roles as defined by the project.

Customer

Customers comprise the business units that identified the need for the product or service the project will develop. Customers can be at all levels of an organization. Because it is frequently not feasible for all the customers to be directly involved in the project, the following roles are identified:

Customer representatives are members of the customer community who are identified and made available to the project for their subject matter expertise. Their responsibility is to accurately represent their business units' needs to the project team and to validate the deliverables that describe the product or service that the project will produce. Customer representatives are also expected to bring information about the project back to the customer community. Toward the end of the project, customer representatives will test the product or service the project is developing, using and evaluating it while providing feedback to the project team.

Customer decision makers are those members of the customer community who have been designated to make project decisions on behalf of major business units that will use or will be affected by the product or service the project will deliver. Customer decision makers are responsible for achieving consensus of their business unit on project issues and outputs and communicating it to the PM. They attend project meetings as requested by the PM, review and approve process deliverables, and provide subject matter expertise to the project team. On some projects, they may also serve as customer representatives or be part of the steering committee.

Stakeholders

Stakeholders are all those groups, units, individuals, or organizations—internal or external to the organization—which are impacted by or can impact the outcomes of the project. This includes the project team, sponsors, steering committee, customers, and customer coworkers who will be affected by the change in customer work practices due to the new product or service; customer managers affected by modified workflows or logistics; customer correspondents affected by the quantity or quality of newly available information; and other similarly affected groups.

Key stakeholders are a subset of stakeholders who, if their support were to be withdrawn, would cause the project to fail.

Vendors or Service Providers

Vendors or service providers are contracted to provide additional products or services the project will require and are additional members of the project team.

Common Project Management Challenges

Oddly, there are a number of common challenges or constraints that PMs deal with on a regular basis. The most famous are called the triple constraints, which are cost, time, and scope.

Cost

All projects have a finite budget; the customer is willing to spend a certain amount of money for delivery of a new product or service. If you reduce the project's cost, you will either have to reduce its scope or increase its time.

Time (Schedule)

As the saying goes, "time is money," a commodity that slips away too easily. Projects have a deadline date for delivery. When you reduce the project's time, you will either have to increase its cost or reduce its scope.

Scope

Many projects fail on this constraint because the scope of the project is either not fully defined or understood from the start. When you increase a project's scope, you will have to increase either its cost or time.

More recently, the triangle has given way to a project management diamond: Cost, time, scope, and quality are now the four vertices with customer expectations as a central theme.

Besides the obvious constraints, most project management text will also agree that the following constraints should be in the mix:

Stakeholder and customer satisfaction
Meeting business case objectives
Customer/end-user adoption
Quality of delivery
Meeting governance criteria
Benefits realization

Managing Project Success

Another area that all three methodologies appear to agree on is the concept of project success. All three agree that success metrics should be established as quickly as possible in the planning phase.

Project success metrics or criteria are the standards by which the project will be judged at the end to decide whether or not it has been successful in the eyes of the customers and other stakeholders.

Successful organizations take the guesswork out of this process: They define what success looks like, so they know when they have achieved it. If a PM wants project success, it must be defined.

Almost as important is how the metric or criteria is documented. This is a good guideline that all three bodies of knowledge would find acceptable:

Name of success criteria
How it is going to be measured?
How often it is going to be measured?
Who is responsible for measuring it?

Measurements are discrete or continuous. Discrete is easy as it simply is a yes/no answer. Was it accomplished or not? Continuous measurement is typically a percentage. For example, 30% of this task was completed. In less finite environments, a Likert scale can be substituted. A Likert scale is the most widely used approach to scaling responses in survey research. As an example, on a scale from one to five, how would you rate the material you are now reading?

There continues to be a bevy of new thoughts about effective project management. For example, the PMBOK is in its sixth edition. However, the most intriguing thing to impact basic project management in several years is the knowledge and ability to apply Lean concepts and Agile techniques. This application makes all projects better, faster, and more cost-effective.

Currently, the Project Management Body of Knowledge (PMBOK®) (Sixth Edition) is the latest version in the market and was released on September 6, 2017. PMBOK contains the entire syllabus that would be covered in the PMP certification examination and is only available for registered users on the PMI Website. According to the history of PMBOK® Guide, PMI® would release an updated version of the PMBOK® Guide every four to five years.

Chapter 2

A Lean History of Lean

A project manager (PM) who would like to apply Lean and Agile concepts should understand the history of both Lean and Agile in addition to the three faces of project management as summarized in Chapter 1. This chapter is designed to give a brief history of Lean as well as Lean Six Sigma (LSS) because both methodologies are used in this book.

Six Sigma was developed by Motorola in 1981 in an effort to reduce defects. During the 1980s, it spread to recognized companies, including General Electric and Allied Signal. Six Sigma incorporated total quality management (TQM) as well as statistical process control (SPC) and expanded from a manufacturing focus to other industries and processes. Motorola documented more than $16 billion in savings. This is when many other companies decided to adopt the methodology. Naturally, the Six Sigma methodology has evolved over time. A core belief is that manufacturing and business processes share characteristics that can be measured, analyzed, improved, and controlled.

In 1988, Motorola won the MBNQA for its Six Sigma program. Six Sigma promotes the following concepts:

Critical to Quality: attributes of the most importance to the customer
Defect: failing to deliver what the customer wants
Process Capability: what the process can deliver
Variation: what the customer sees and feels
Stable operations: ensuring consistent, predictable processes to improve what the customer sees and feels
Design for Six Sigma: designing to meet customer needs and process capability

In 1999, GE reported $2 billion in savings attributable to Six Sigma in its 2001 annual report. It discussed the completion of more than 6,000 Six Sigma projects and the probability of yielding more than $3 billion in savings by conservative estimates. Other early adopters of Six Sigma include the following:

- Bank of America
- Bechtel
- Borusan
- Brunswick Corporation
- DuPont
- EDS
- Honeywell
- Idex
- Raytheon
- Shaw Industries
- Smith and Nephew
- Starwood (Westin, Sheraton, Meridian)
- Wildcard Systems

Sigma

Sigma is a statistical measurement of variability, showing how much variation exists from a statistical average. Sigma basically measures how far observed data deviate from the mean or average.

The term Six Sigma is a statistical measurement based on defects per million opportunities (DPMO). A defect is defined as any nonconformance of quality. At Six Sigma, only 3.4 DMPO may occur. In order to use sigma as a measurement, there must be something to count and everyone must agree on what constitutes a defect. Normal distribution models generally explore only Three Sigma, which is essentially 6,210 DPMO.

It is important to know that many processes are acceptable at lower sigma levels. Six Sigma is considered perfect but may not be cost-effective or practical.

Sigma (σ) is a symbol from the Greek alphabet that is used in statistics when measuring variability. In the Six Sigma methodology, a company's performance is measured by the sigma level. Sigma levels are a measurement of error rates. It costs money to fix errors, so saving this expense can be directly transferred to the bottom line.

Popular tools taught in Six Sigma include, but are not limited to, the following:

Affinity Diagram or Kawakita Jiro (KJ) method
Control Plan
Critical Path Analysis
Failure mode and effects criticality analysis (FMECA)
Failure mode effects analysis (FMEA)
Histogram
Ishikawa diagram or fishbone
Measurement System Analysis (MSA)
Pareto Chart
Process Mapping
Quality Function Deployment (QFD), also known as House of Quality
Scatter Diagram
Supplier-Input-Process-Output-Customer (SIPOC) analysis

Lean manufacturing is a production practice that concentrates on the elimination of waste. It is based on the total production system (TPS), introduced originally by Toyota, and is also based on the principals of TQM. TQM capitalizes on the involvement of management, workforce, suppliers, and even customers in order to meet or exceed customer expectations.

Originally, Lean identified the following as the worst forms of waste:

Transportation
Inventory (all components: work in progress and finished product not
   being processed)
Movement
Waiting
Overproduction
Over processing
Defects
Skills

An easy way to remember the primary forms of waste is T-I-M W-O-O-D-S.

Eventually, Lean evolved to consider additional types of waste. Lean thinking is designed to do the following:

Shrink lead times
Save turnover expenses

Reduce setup times
Avoid unnecessary expenses
Increase profits

Lean focuses on getting the right things to the right place at the right time in the right quantity while minimizing waste. Lean also makes the work simple enough to understand, to do, and to manage. The very nature of Lean would suggest that it would be wasteful to spend time trying to understand manuals or complicated processes, so it is best to simplify language.

Typical tools promoted in Lean include the following:

5S
Error proofing
Current reality trees
Conflict resolution diagram
Future reality diagram
Inventory turnover rate
Just in time (JIT) theories
Kaizen
Kanban
Lean metric
One-piece flow
Overall equipment effectiveness
Prerequisite tree
Process route table
Quick changeover
Standard rate or work
Takt time
Theory of constraints
Total productive maintenance
Toyota production system
Transition tree
Value added to non-value added lead time ratio
Value stream mapping
Value stream costing
Visual management
Workflow diagram

Another variation of Lean is Lean office. There are seven primary principles to a Lean office. These are continually being updated.

1. Committed leadership
2. Establishing metrics and goals
3. Standardized processes
4. 5S—a physical organizational system
5. Minimal work in progress (WIP)
6. Positive workflow
7. An understanding of demand

Lean offices use a daily management system and communication similar to Agile. They are typically visual with signs pointing the way and labels on cabinets. Finally, it fosters a culture of continuous improvement.

LSS began in the late 1990s. Both Six Sigma and Lean already started expanding to include service as well as manufacturing. Allied Signal and Maytag independently started experimenting with using both methodologies. Employees were cross-trained. Maytag was the first to recognize that Lean manufacturing and Six Sigma do not conflict with each other and, in fact, are perfect complements. This thinking is shared by the U.S. Army. Everyone involved in the movement to implement LSS saw the power of combining both toolboxes and attacking both defects and wastes. There were also many shared tools that are apparent in both methodologies that reduce the learning curve.

LSS can be used in any industry including finance, construction, government, health care, insurance, and hospitality. LSS is about increasing quality and profit. The new tools include methodologies based on teamwork as a principle. Process improvement is not a linear process in which each component is handed off to another department or individual. Each member of the process is involved in improving client satisfaction.

The new leaner tools focus on continuous improvement as a guiding principle. The road to quality is paved with small incremental improvements. Major sweeping changes seldom work. As this country moves its business style from control to management to leadership, we are finding that the people actually doing the work are the most capable of identifying changes necessary to improve quality. Leadership must listen and implement changes rather than direct the solutions. Some examples are the following:

Improving forecast accuracy
Reducing volume of rejected orders
Improving consumer loan cycle time
Reducing engine installation times
Eliminating mistakes in an operating room

Reducing pharmacy dispensing error rates
Improving the effectiveness of employee hazard recognition
Reducing process variation costs related to manufacturing

Before examining LSS, the topic of continuous improvement (CI) programs should be explored. Most companies have some sort of quality control program. These programs may be formal or informal. Some programs have defined documents and manuals, and other quality programs are not actually recorded or tracked. Quality is obviously a large piece of CI. All CI programs ask two questions: (1) Who are the customers? (2) What will it take to satisfy them?

Both Lean and Six Sigma endorse the plan, do, check, act (PDCA) model. This popular project management tool is easy to understand. It is also called the Deming wheel or Deming cycle.

Plan: Identify an opportunity and plan for change.
Do: Implement the change on a small scale.
Check: Use data to analyze the results of the change and determine whether it made a difference.
Act: If the change was successful, implement it on a wider scale and continuously assess the results. If the change did not work, begin the cycle again.

Both Lean and Six Sigma support the idea of CI. It is an ongoing effort to improve products, services, or processes. It can be incremental improvement (over time) or breakthrough improvement (all at once). CI programs often are not proactive and are presented with a problem up front. Within any problem-solving model, there are four steps to remember: define the problem, generate the solution, evaluate and select an alternative, and implement.

LSS uses a set of quality tools that are often used in TQM. These tools, sometimes referred to as problem-solving tools, include the following:

Control charts
Pareto diagrams
Process mapping
Root cause analysis
SPC

As noted continually throughout this book, LSS largely depends on the Define, Measure, Analyze, Improve, and Control (DMAIC) model that was developed by Motorola and later enhanced by General Electric.

> ***Define*** the process improvement goals that are consistent with customer demands and enterprise strategy.
>
> ***Measure*** the current process and collect relevant data for future comparison.
>
> ***Analyze*** the relationships of all the factors including the variations.
>
> ***Improve*** or optimize the process based upon the analysis.
>
> ***Control*** to ensure that any variances are corrected before they result in defects.

LSS uses the tools above. It also uses the martial arts designations—white, yellow, green, black, and master black belts—to denote the level of expertise. The generally accepted belt designations are as follows:

> Yellow belt (YB): individual trained in the basic application of Six Sigma management tools
>
> Green belt (GB): individual who handles LSS implementation along with other regular job responsibilities
>
> Black belt (BB): individual who may devote 100% of his or her time to LSS initiatives
>
> Master black belt (MBB): individual who acts in a teaching, mentoring, and coaching role

Lean Six Sigma Yellow Belt

Although there are inconsistencies in training and certification of LSS Yellow Belts (LSS-YB), most training providers will agree that LSS-YB training is designed to provide students with a strong understanding of the core Lean Six Sigma methods. Programs and certifications are designed to prepare the LSS-YB student to participate as contributors to LSS projects and perform simple improvements in their own work area. Lean Six Sigma Yellow Belts are not typically experts in the methodology. The goal of LSS-YB training and certification is to understand the activities, deliverables, and key concepts of problem solving.

The Lean Six Sigma Yellow Belt graduate is expected to improve processes in their own day-to-day work product and support the efforts of Lean
Six Sigma Green Belts.

(See Appendix C: Lean Six Sigma Competency Models and Job
Descriptions: Yellow Belt.)

Lean Six Sigma Green Belt

Lean Six Sigma Green Belt (LSS-GB) training and certification is more
defined than LSS YB programs. A certain set of competencies is required.
(See Appendix: Competency Models.) Typically, the LSS-GB may perform/
assist in the following activities.

Provide Lean Six Sigma Black Belts with another resource for project
completion
Implement/Execute process improvement projects
Incorporate Lean Six Sigma method into all aspects of the project Provide
Lean Six Sigma Charts and Graphs
(See Appendix C: Lean Six Sigma Competency Models and Job
Descriptions: Green Belt.)

Lean Six Sigma Black Belt

The competencies of a Lean Six Sigma Black Belt (LSS-BB) are more in-
depth because LSS-BB may be spending up to 100% of their time on LSS
projects. Although, there is also the tendency for the LSS-BB to have an
over-all higher formal business or quality education than a Lean Six Sigma
Green Belt. A Lean Six Sigma Black Belt generally also has some subject
matter expertise in a particular area such as manufacturing, information
technology, or finance.

Core competencies for the LSS-BB, include, but are not limited to, the following abilities to:

Determine how often something happens or is observed
Determine if two or more things are different from each other
Determine if there is a relationship between two or more things
Determine when and/or how something must be adjusted

Determine if there is a pattern in something across time
Determine the best operating conditions for something
Determine which things cause other things to happen

(See Appendix C: Lean Six Sigma Competency Models and Job Descriptions: LSS Black Belt.)

Lean Six Sigma Master Black Belts

The Lean Six Sigma Master Black Belt (LSS-MBB) mostly functions in the role of coach, teacher, or mentor. Often companies will employ an LSS-MBB to build an internal LSS program.

Training focuses a little more on the history and concepts in order to develop critical thinking skills. For example, topics such as planning, or development and enterprise-wide deployment are discussed. Statistical and analytical tools are shown from the perspective of how to teach and demonstrate. (See Appendix C: Lean Six Sigma Competency Models: Lean Six Sigma Master Black Belt.)

The roles, responsibilities, and education for LSS-GBs and LSS-BBs are the most consistent. The major difference between the LSS-GB and LSS-BB is that LSS-GBs have a regular job where they apply process improvement, whereas many times the LSS-BB are engaged solely in process improvement efforts. And, these efforts may be outside the area of the LSS-BB subject matter expertise. For example, the LSS-Green Belt might not leave the manufacturing floor or the finance office. The LSS-Black Belt could be a subject matter expert in information technology but could be asked to manage a project in an entirely different department such as marketing or human resources.

To be a successful LSS-GB, individual expertise should be blended with LSS tools and methods. To be a successful LSS-BB or LSS-MBB requires an overall understanding of business. This knowledge can be gained from a Master of Business Administration (MBA) degree, or a project management/quality engineering background. These days since information technologists have to have a strong understanding of how departments interface with one another, they are often strong candidates for LSS-BB or LSS-MBB certification. Likewise, individuals who have owned a business of any size where they had to understand and or manage all the various aspects of a business do well.

Other roles in the LSS organization include the sponsor, process owner, and cross-functional teams. The sponsor is generally the person paying for the project. The process owner is the person normally responsible for process success, and the cross-functional team is the ideal team promoted by LSS—a team made up of multiple disciplines to include functional expertise, finance, marketing, and operations.

The roles and responsibilities in LSS are still rooted in Total Quality Management (TQM). In a TQM effort, all members of an organization participate in improving processes, products, and services. TQM practices are based on cross-functional product design and process management. Other components related to LSS also covered in TQM include the following:

Committed leadership
Customer and employee involvement
Feedback
Information
Overall quality management
Strategic planning
Supplier relations

Five Laws of Lean

The five laws have evolved over time and now are commonly referred to as the five laws of Lean Six Sigma. They are a collection of key ideas derived both from Lean manufacturing and Six Sigma. These laws are a quick way to sum up the philosophy of Lean. They appear in various lengths of text, but this is the most simplified version we have been able to locate that does not take away from the overall meaning of the laws.

First it is important to explain that the concept of the zeroth law appears in other bodies of knowledge such as thermodynamics and robotics. It was adopted by Lean and Lean Six Sigma.

1. The zeroth law: The first law is called so because all other principles are built upon this fundamental one. It states, "The law of the market—customer critical to quality defines quality and is the highest priority for improvement, followed by ROI (Return on investment) and net present value."
2. The first law: This is called as the law of flexibility. It states, "The velocity of any process is proportional to the flexibility of the process."

Interpretation: The more the process is receptive and flexible to adopt changes, the better the progress of the project implementation is.

3. The second law: The second law is known as the law of focus; it is defined as "20% of the activities in a process cause 80% of the delay." This can be interpreted as the main causes of delay of activities originated from just 20% of activities, which thus enables a faster refocus during the reorientation phase.

4. The third law: The law of velocity, as the third law is known, is stated as "the velocity of any process is inversely proportional to the amount of WIP." This is also called "Little's law." This explains how the inertia of WIP bears heavily on the velocity of project implementation. The higher the number of WIPs (read: unfinished tasks), the lower the speed of progress due to various ground-level handicaps.

5. The fourth law: The fourth law, which is the last of the five laws of LSS, is defined as "the complexity of the service or product offering adds more non-value, costs, and WIP than either poor quality (low sigma) or slow speed (not lean) process problems." The bulky nature of products is against the foundation of Lean manufacturing principles. The bulk, complex manufacturing process and product and service specifications contribute to rendering the offerings redundant. As an illustration of this fourth law of LSS, you can try and reason out why passenger cars are more and more becoming driver-friendly despite their complex engineering features and functions.

Project managers interested in gaining a better understanding of the history of process improvement will benefit from studying material first presented within the TQM framework. Works by W. Edwards Deming and Joseph Juran are still prevalent today. Other major authors include Kaoru Ishikawa, A. V. Feigenbaum, and Philip B. Crosby.

This book uses the term Lean thinking. Lean thinking was devised by James P. Womack and Daniel T. Jones to explain the philosophy they used when studying processes and solutions at Toyota. Lean thinking is a way of thinking about activities and identifying the potential or inadvertent waste. Their philosophy focused on the following concepts:

Flow
Push/pull
Value
Value Stream Mapping (VSM)

Flow

If flow were a person instead of a concept, his or her goal would be to move products through a production system without separating things into lots. Primarily a manufacturing term, over the years, it has morphed into the service industry. Lean is primarily concerned with eliminating waste and improving flow by following the Lean principles and a defined approach to implementing each of these principles.

As a side note, Six Sigma methodology is focused on reducing variation and improving process yield by following a problem-solving approach using statistical tools. This is why Lean and Six Sigma often work together.

Push/Pull

Push/pull, unlike flow, has not made as successful a transition from manufacturing to include service.

Supply chain management (SCM) is to create a solution, that is, supply, for a goal or issue, that is, demand. Supply chain models of push type and pull type are opposite in terms of a demand and supply relationship. Push type is represented by make to stock (MTS) in which the production is not based on actual demand, and pull type is represented by make to order (MTO) in which the production is based on actual demand.

One of the major reasons why SCM currently receives so much attention is that information technology enables the shifting of a production and sales business model from push type to pull type. Pull-type SCM is based on the demand side, such as JIT and a continuous replenishment program (CRP) or actual demand assigned to later processes.

Lean thinking is a movement of practitioners who experiment and learn in different industries and conditions to Lean think any new activity. It relies heavily on social innovations. Social innovation focuses on the process of innovation and how innovation and change take shape. Social innovation focuses on new work and new forms of cooperation (business models), especially those that work toward a sustainable society.

There are many technical tools used by Lean professionals. But it is the people factor that separates Lean thinking the most from other problem-solving methodologies, which aligns well with Agile project management.

Value

Value in Lean simply stated is the value the customer finds in your product or service. There are many tools associated with concept of "value."

Value Stream Mapping

Lean value streams analyze the current state and design a future state for a series of events that take a product or service from its beginning through to the customer. Although not necessary, mapping usually employs standard symbols to represent items and processes. So, to correctly interpret the diagram, the PM needs to be somewhat familiar with the representations. The symbols are different from those used in flowcharting. Some are intuitive, and others rely on standard pictures used to depict certain items or situations related to JIT manufacturing.

Although value stream mapping is often associated with manufacturing, it is also used in logistics, supply chain, service-related industries, and health care.

A closely related concept is Business Process Mapping (BPM). BPM is used to assist organizations in becoming more efficient. A clear and detailed business process map or diagram allows outside firms to come in and look at whether or not improvements can be made to the current process.

BPM takes a specific objective and helps to measure and compare that objective alongside the entire organization's objectives to make sure that all processes are aligned with the company's values and capabilities.

Most experts will say that Value Stream Mapping (VSM) is a subset of BPM, but some disagree and think of VSM as a stand-alone activity. Lean is a production practice with the key tenet of preserving value with less work. Operations that fail to create value for the end customer are deemed wasteful. Eliminating waste and superfluous processes reduces production time and costs. This is the objective for VSM.

VSM immediate benefits relate to productivity, error reduction, and customer lead times. BPM is typically looking for longer-term benefits including improvements to financial performance, customer satisfaction, and employee morale. By default, VSM often accomplishes this as well.

Project managers benefit by understanding the concept of Lean primarily because it is a methodology dedicated to reducing waste and making project faster or more efficient. Many of the tools can be used in Agile Sprints—discussed later in this book—to make the Sprint even faster and the identification of non-value activities easier.

Chapter 3

Agile Comprehensive with an Emphasis on Scrum

The genesis of Agile is found in a group of software development methodologies. However, many of these tools and techniques may be applied to projects in general. Agile methodology is based on iterative development with which requirements and solutions evolve through collaborative effort. Agile supports self-organizing, cross-functional teams. It isn't surprising that this morphed into a structure for the project manager (PM).

Agility has emerged as the successor to mass production. It is a comprehensive response to the business challenges regarding rapid growth of profiting from rapidly changing and continually fragmenting global markets for high-quality, high-performance, customer-configured goods and services. It is a continual readiness to change, sometimes radically, what companies and people must do and how they will do it.

The approach to Agile Project Management is foreign to traditional project management administration. It is possible, however, for the PM to benefit from Agile techniques even if the project itself has not been labeled Agile.

Many of the tools and methods can easily be incorporated into certain projects. This would include military project management even though, at first glance, this would not seem possible.

It is important to know a little about the history and evolution of Agile. This is valuable even though this chapter mostly covers Scrum, a form of Agile management, as opposed to the other things the Agile umbrella covers.

Throughout this book, tools project managers can use from the Agile toolbox are referred to as Agile techniques. Likewise, this chapter is dedicated to understanding Agile as a stand-alone methodology.

Agile theory is based on an Agile manifesto. Sometimes, a manifesto is confused with the term body of knowledge (BOK). Whereas there are some synergies between what we consider a BOK as presented in other disciplines, such as project management, a manifesto is more about purpose than technicalities.

The Agile manifesto was written in February 2001 at a summit of 17 independent-minded practitioners. Most of the participants had a programming background. According to the Agile Alliance, many participants had different ideas of what constituted Agile theory. However, they did agree on four main values:

> Individuals and interactions over processes and tools
> Working software over comprehensive documentation
> Customer collaboration over contract negotiation
> Responding to change over following a plan

These four values were intended to supplement the Agile 12 principles. History on how these 12 principles were developed originally remains a bit fuzzy. Still, these items provide a better understanding of the framework and intention of Agile methods. The principles are defined as follows:

1. Our highest priority is to satisfy the customer through early and continuous delivery of valuable software.
2. We welcome changing requirements even late in development. Agile processes harness change for the customer's competitive advantage.
3. Deliver working software frequently—from a couple of weeks to a couple of months with a preference for the shorter time scale.
4. Businesspeople and developers must work together daily throughout the project.
5. Build projects around motivated individuals. Give them the environment and support they need and trust them to get the job done.
6. The most efficient and effective method of conveying information to and within a development team is face-to-face conversation.
7. Working software is the primary measure of progress.
8. Agile processes promote sustainable development. The sponsors, developers, and users should be able to maintain a constant pace indefinitely.

9. Continuous attention to technical excellence and good design enhances agility.
10. Simplicity—the art of maximizing the amount of work not done—is essential.
11. The best architectures, requirements, and designs emerge from self-organizing teams.
12. At regular intervals, the team reflects on how to become more effective, then tunes and adjusts its behavior accordingly.

Lean, as addressed in the previous chapter, does not technically have a manifesto or a specific industry agreed-upon list of principles. However, much of Lean thinking is based on W. Edwards Deming's 14 points from his book *Out of the Crisis*. These points provide for a good comparison between Lean thinking and Agile techniques as Agile supports many of Dr. Deming's unprecedented points. Dr. Deming taught that most quality issues were systemic (process-related) and, therefore, the responsibility of management. Agile shares most of these philosophies. Summarized, these points are as follows:

1. Create constancy of purpose for improving products and services.
2. Adopt the new philosophy.
3. Cease dependence on inspection to achieve quality.
4. End the practice of awarding business on price alone; instead, minimize total cost by working with a single supplier.
5. Improve constantly and forever every process for planning, production, and service.
6. Institute training on the job.
7. Adopt and institute leadership.
8. Drive out fear.
9. Break down barriers between staff areas.
10. Eliminate slogans, exhortations, and targets for the workforce.
11. Eliminate numerical quotas for the workforce and numerical goals for management.
12. Remove barriers that rob people of pride of workmanship and eliminate the annual rating or merit system.
13. Institute a vigorous program of education and self-improvement for everyone.
14. Put everybody in the company to work accomplishing the transformation.

Earned Value Management

Even though Agile has a flexible attitude about change that sometimes leaves the traditional PM unsettled, Agile excels at cost management. Earned value management (EVM) is a familiar term to most PMs, but Agile takes it one step past the typical formula and makes full use of these terms:

EV: Budgeted cost of work performed (BCWP)
PV: Budgeted cost of work scheduled (BCWS)
AC: Actual cost of work performed
BAC: Budget at completion of project
SV: Schedule variance (EV − PV)
CV: Cost variance (EV − AC)
SPI: Schedule performance index: EV/PV (≥1 means project schedule under control)
CPI: Cost performance index (EV/AC) (≥1 means project cost under control)
EAC: Estimate at completion or expected total cost of the project at its end (EAC = BAC/CPI)
ETC: Estimate to complete or expected additional cost needed to complete the project from the point of calculation (ETC = EAC − AC)

Recognized Certifications in Agile

How individuals define the BOK for Agile is confusing. Generally, any type of legitimate certification is based on a BOK. However, Agile has many methodologies, such as Scrum, XP, Lean, FDD, Crustal, DSDM, and draws from Just-in-Time and Kanban philosophies.

Proving a comprehensive BOK, covering all the ideas and concepts that drive Agile, is not yet possible. This means that a number of special certifications exist through private vendors. However, the most respected and standardized one is that provided by the Project Management Institute (PMI).

The PMI offers a PMI Agile certified practitioner (PMI-ACP)® designed to formally recognize an individual's knowledge of Agile principles and skills with Agile techniques.

The prerequisites to sit for this exam include, but may not be limited to, the following:

2,000 hours of general project experience working on teams. A current PMP® or PgMP® will satisfy this requirement but is not required to apply for the PMI-ACP.

1,500 hours working on Agile project teams or with Agile methodologies.
   This requirement is in addition to the 2,000 hours of general project
   experience.
21 contact hours of training in Agile practices.

Other recognized entities that provide well-respected certifications in Agile
include, but are not limited to, the following:

Agile Alliance: The Agile Alliance is the original global Agile community
   with a mission to help advance Agile principles and practices regardless
   of methodology.
Scrum Alliance: The Scrum Alliance is a nonprofit professional member-
   ship organization that promotes understanding and usage of Scrum. The
   Scrum Alliance offers a number of professional certifications:
   Certified Scrum Master (CSM)
   Certified Scrum Product Owner (CSPO)
   Certified Scrum Developer (CSD)
   Certified Scrum Professional (CSP)
   Certified Scrum Coach (CSC)
   Certified Scrum Trainer (CST)
Platinum Edge: Providers of training classes worldwide and also develop-
   ers of transition strategies and coaching for organizations moving to
   Agile project management.
Scaled Agile Framework® (SAFe®): Designed to empower complex orga-
   nizations to achieve the benefits of Lean-Agile software and systems
   development at scale.
Disciplined Agile (DA): A toolkit that provides straightforward guidance
   to help organizations choose their way of working (WoW) in a context-
   sensitive manner, providing a solid foundation for business agility.
   This product has been recently adopted by the Project Management
   Institute.

Agile Basic Tools and Techniques

Probably the most recognized word when Agile is applied to project management is Scrum. Scrum is an Agile methodology that can
be applied to nearly any project. Although originally used for software development, Scrum theory is now used in all types of business
ventures.

Scrum

The Scrum model advises that each sprint begins with a brief planning meeting and closes with a review. Scrum is a management and control process that cuts through complexity and is a simple framework for effective team collaboration.

Starting with a basic daily Scrum, there are three questions the PM asks:

1. What did you do yesterday?
2. What will you do today?
3. Are there any impediments in your way?

The Scrum team, ideally, includes everyone who touches the project. These meetings are held in an area where everyone can stand and face each other, generally in a circle. The purpose of team players standing is to promote that the session should be quick and precise. Scrum sessions are held for individual projects. However, the Scrum meeting could support a daily report as well as reports on various projects.

By focusing on what each person accomplished yesterday and will accomplish today, the team gains an excellent understanding of what work has been done and what work remains to be done. Sometimes to make this easier, the Scrum master may refer to a burn down chart (see Figure 3.1). However, this chart may also be used to show the progress of a sprint as discussed in the next section.

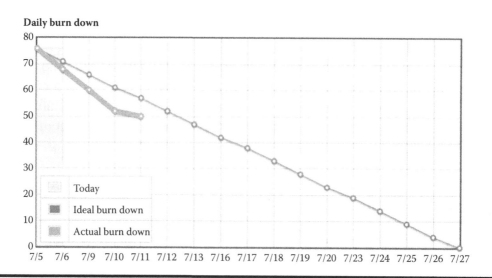

Figure 3.1   Burn down chart.

The Scrum master does anything possible to help the team perform at its highest level. This involves removing any impediments to progress, facilitating meetings, and doing things like working with the product owner to make sure the product backlog is in good shape.

Sprints

In the Scrum version of Agile, sprints are collections of work confined to a regular, repeatable work cycle, known as a sprint or iteration. These iterations can be anywhere from 1 week to 30 days but should be the same duration. This allows for less to remember about the sprint schedule, and the planning becomes more accurate. During this time, the Scrum team works on very specific and agreed-upon work. Nothing can be changed during the sprint.

Step One is a sprint planning session. Everyone who touches the process should be involved. Whereas daily Scrum meetings typically include those actually doing the work and who will report to other interested parties if appropriate, for a sprint to be successful, everyone should be involved. Naturally, this would be when a consensus should be reached about the sprint duration. The optimum sprint duration depends on many factors that include, but should not be limited to, availability of resources and urgency of the project.

Step Two is to decide what piece of the backlog should be tackled first. Most sprints include a little more than can be achieved, which is why some employees are not as comfortable as others with the process.

Once the body of work is agreed upon, tasks are taken one by one, in logical order, and the objectives of the first sprint are determined.

Sprint Retrospectives

Sprint retrospectives are meetings at the end of each sprint in which the Scrum team discusses what went well, what could change, and how to make any changes. Typical questions for discussion include the following:

What went well during the sprint cycle?
What went wrong during the sprint cycle?
What could we do differently to improve?

Agile Stages

It may be easier to think of the Agile process in terms of steps or stages once the terminology is understood.

> Stage One: The product or process owner identifies the vision.
> Stage Two: The Scrum master or product/process owner creates the roadmap.

The roadmap is the high-level view.

> Stage Three: The Scrum master or the product/process owner releases a plan identifying the timetable.
> Stage Four: Product/process owner, the Scrum master, and the Scrum team plan the sprints. Sprint planning takes place at the beginning of each sprint.
> Stage Five: During each sprint there are daily Scrum meetings that last no longer than 15 minutes.
> Stage Six: At the end of every sprint, a report is given to the product/process stakeholders.
> Stage Seven: The team holds a sprint retrospective.

Agile Manufacturing

The term Agile manufacturing was watermarked by Francois de Villiers in his work *Lean and Agile World Class Manufacturing*. Villiers mentions that his work was never meant to be published. He said that he compiled the manual as a personal self-help text. It did, however, evolve into this sophisticated account of both Lean and Agile manufacturing.

Villiers describes Agile manufacturing as tools, techniques, and initiatives that enable a plant or company to thrive under conditions of unpredictable change. Agile manufacturing not only enables a plant to achieve rapid response to customer needs, but it also includes the ability to quickly reconfigure operations—and strategic alliances—to respond rapidly to unforeseen shifts in the marketplace. In some instances, it also incorporates mass customization concepts to satisfy unique customer requirements. In broad terms, it includes the ability to react quickly to technical or environmental surprises. It is a means of thriving in an environment of continuous change by managing complex inter- and intrafirm relationships through innovations in technology, information, communication, organizational redesign, and new marketing strategies.

Agile Change Management

In an Agile project, the authority to approve many changes is delegated to the roles in the Agile team. PRINCE2, as noted in Chapter 1, has a specific change management plan that is a strong basis for other concepts. Other project models, however, align closely with Agile. The specific differences are as follows:

The product owner and PM have more authority to approve change. Minor changes are just done.
Requirement trade-off is used to allow larger changes and still keep the project within tolerance.

Agile Project Management

Agile project management promotes a value-driven approach that allows the PM to deliver high-priority, high-quality work. It is an iterative, incremental method of managing the design and building activities. It works well in engineering and information technology.

Agile methods are mentioned in the Guide to the Project Management Body of Knowledge (PMBOK® Guide) under the project life cycle definition and are referred to as an adaptive project life cycle. Typically, adaptive life cycles are iterative and incremental. The caveat is that the iterations in Agile processes are rapid—two to three weeks in length—with fixed resources.

One of the social innovations of Agile is that it aligns well with performance management theory.

Performance management is a science that takes the emphasis in organizations away from command and control toward a facilitation model of leadership. It examines not only the compensation of the individual, but items such as the availability of tools necessary to be successful (reporting and tracking structures) and, in some cases, even environmental issues, such as lighting, temperature, and location of the facility.

Work in this century has become very individualized. Project managers must consider the challenge of managing virtual teams and individuals who telecommute. Agile offers a solid return to teams and uses collective intelligence.

Project managers must often rely on their personal influence to gain buy-in and productivity rather than relying on human resources policies.

Employers who include disability issues in corporate diversity policies enrich and enhance workplace benefits in the new economy. However, the PM is now charged with the responsibility of committing and implementing these policies. It is up to managers to design disability-friendly strategies for the workplace, which often includes education of the staff.

Being a compliance officer, employee coach, and performance manager has added new stress to the job of the PM in addition to management duties, such as resource leveling, finance, and marketing. The information provided in the following chapters is designed to create a balance and to provide resource information.

Agile Challenges

As with any kind of change, Agile project management is subject to challenges. Because Agile is, by nature, team-oriented, and teams are often geographically distributed, most of the trials and tribulations are apparent in this area.

Agile process development is becoming more mainstream as teams are migrating from Waterfall to Agile development. Because this shift does not always give the team a clear roadmap for the next step, some team members become frustrated, and some managers just want to see to a traditional plan.

A Lean enterprise operates by creating products and services to meet customer orders rather than marketing forecasts. An Agile enterprise aggressively embraces that type of thinking but is a little better at making the shift in midstream.

To be Agile is to be capable of operating profitably in a highly competitive environment. This environment is continually changing and unpredictable. Agile techniques allow the PM to move quickly from decision making to action and innovation. Whereas much of Lean thinking is dedicated to speed and eliminating wasteful activities, Agile concentrates on how to work effectively in uncertain circumstances in which the direction is determined by the complexity of the product or service as opposed to a particular protocol.

In summary, since the purpose of this chapter is to cover Agile as it applies to the project manager, we will close with a set of terms and concepts related to Agile project management. The project manager should keep in mind that not all projects are good candidates for Agile or more specifically the Scrum method of management. Some project must remain Waterfall due to compliance or, at least temporarily, to company culture and understanding. As a reminder, Waterfall, or more commonly called the Waterfall method, is a traditional way of organizing projects in a systematic way.

Examples of Waterfall include the Six Sigma version of DMAIC as well as scientific problems solving or the System Development Life Cycle (SDLC).

This is a summary of the ideas and concepts related Agile project management that are useful to project managers trying to quickly digest Agile as it relates to Scrum, which again, is the way Agile projects are often managed.

Backlog—a prioritized list of all the things necessary to complete the project
Sprint—a period of work (generally two to four weeks)
Promise—what is intended to be delivered in that particular sprint
Scrum—a framework to organize the project

Scrum Master—a project manager who helps the team stay organized, motivated, and makes sure the information/resources/tools needed are available
Stand-up Meeting—a meeting held at the same time every day where brief reports are provided on the daily status for each team member usually including the following: (a) what got done; (b) what tasks are scheduled; and (c) what potential obstacles are in the way.

Stories—what the customer wants and why
Timebox—maximum time allotted to produce something valuable to the customer

In many Agile environments a Scrum board often called a Kanban board. A Scrum board is a tool that helps teams make sprint backlog items visible. There are different ideas of how a Scrum board should look and software available to build and maintain the board. Whichever visual form it takes, the board is updated by the team frequently and is designed to shows all items that need to be completed for the current sprint. Most Scrum teams use a Scrum board to help organize tasks and track each task through its life cycle. A Scrum board will always include columns for Story, To-Do, Work in Progress (WIP), and Done.

Chapter 4

Initiating the Project

"Begin at the beginning," the King said, very gravely, "and go on till you come to the end: then stop."

Lewis Carroll, *Alice's Adventures in Wonderland*

Before a project can be initiated, a project must be chosen. This is the beginning. However, frequently, the project manager (PM) is not involved in this process. A PM is simply given the assignment. Whereas this chapter assumes that the PM is taking an active role in the selection process, a PM who has been delegated a project will still benefit from the information presented.

In Lewis Carroll's *Alice's Adventures in Wonderland*, the Cheshire Cat tells Alice, "If you don't know where you are going, any road will get you there."

One of the purposes of project management is to create a project purpose as well as a roadmap. PMs should know where they are going and how to get there. However, it often seems that those creating the project have not taken the time to completely consider the culture, environment, or resources necessary to complete the project successfully. Additional challenges exist when the project is not properly aligned with the business goals and objectives.

Any project should be carefully vetted to determine whether or not it benefits the organization. How does this project fit with the overall mission of the company? If a connection cannot be made, it won't matter if traditional project management uses Lean thinking or Agile techniques (Lean and Agile).

Fortunately, PMs can adopt components of Lean and Agile when making a project selection. In traditional project management, a PM is often programmed to consider two types of thinking.

The first is the mathematical method (mathematical) sometimes referred to as constrained optimization. This method involves the calculation of several different mathematical factors.

The second method is the benefit measurement method (benefit). This method enables the PM to effectively compare the benefits and values of one project against another.

Both the mathematical and benefit approaches are effective; however, these methodologies are more robust when enhanced by certain aspects of Lean and Agile.

For example, in Lean, it is not uncommon to select a project by performing a pilot or running an idea hypothetically through the plan-do-check-act (PDCA) model. A PM can perform a type of what-if scenario using the PDCA. This will allow some smaller projects or ideas to be jump-started and others to be killed without much time investment.

Lean projects also consider factors not always apparent in the mathematical or benefit structure, such as the following:

How will the project/process strategically impact the business objectives and outcomes?
How can the project/process be simplified?
How will the project/process align with customer issues and complaints?
How will the project/process handle resources both human and nonhuman?
How will the project/process enhance the core competency of the business?

Some of these questions are answered by default in the Agile approach. This is because during Scrum and sprint sessions, employees are actually speaking to one another and capitalizing on collective intelligence. Both the Scrum and sprint sessions can inadvertently uncover future worthwhile projects to be considered at a later date.

In addition, Agile works with a framework that is not always strongly represented in traditional project management. Agile project selection considers a number of factors that include the following:

Organizational culture
Management philosophy

Timelines
Project breakdown
Roles
Ability to secure an effective cross-functional team
Experience needed
Delivery method

When considering project selection, another methodology, in addition to Lean and Agile, can be used to enhance the traditional mathematical and benefit tactic. This method is Lean Six Sigma (LSS). Lean is an independent methodology, but for more than a decade, it has been partnered with Six Sigma in many environments. Although it can be argued that LSS has far more Six Sigma methodology than Lean influence, together the two methodologies can often give a PM the insight and tools required to conquer more complex problems.

The Six Sigma model respects and uses the PDCA when appropriate but primarily depends on the Define, Measure, Analyze, Improve, and Control (DMAIC) methodology.

Project Selection Using the Lean Six Sigma Method

The primary model in LSS is the DMAIC model. In Six Sigma, the DMAIC is considered a waterfall technique, meaning one phase must be completed before the next phase begins. This is because the Six Sigma practitioner is usually working on large projects. Therefore, different phases of the model may be handled by different departments. In the LSS environment, the PM has hands-on responsibility for most aspects of the project and can approach the model more like a wheel with the flexibility of moving back and forth. It should be noted that this is not the objective, but it does offer more elasticity for the PM.

The first phase of the model, Define, is useful in both project initiation and project planning. Combining certain features of the Lean and Six Sigma methodologies is particularly advantageous in project selection for the PM. Define offers tools that are useful in project selection. These tools are often also referred to as LSS tools.

During the Define phase, a team and its sponsors reach agreement on what the project is and what it should accomplish. Presuming that a draft of the project charter is already in place, the main work in the Define phase is for the project team to complete an analysis of what the project should accomplish and confirm that understanding with the sponsor(s).

Ideally, the team should agree on the problem, which customers are affected, and how the current process or outcomes fail to meet the customer needs through the voice of the customer (VOC) or critical to quality (CTQ).

The terms VOC and CTQ are primarily Lean terms; however, as noted, they are used in the Six Sigma methodology and often are theoretically adopted by Agile.

The outcome of the Define phase includes the following:

A stakeholder's analysis
A high-level map of the processes
CTQ factors
CTQ tree
Supplier, input, process, output, customer (SIPOC)
VOC
Affinity diagram
Strengths, weaknesses, opportunities, threats (SWOT) analysis

Although the Six Sigma methodology promotes Define (DMAIC), rather than plan (PDCA), most of the tools used in Define, currently, were Lean to begin with or heavily rooted in Lean thinking.

Following is a brief description of the above-mentioned Lean and Six Sigma tools.

Stakeholder Analysis

A DMAIC project will require a fundamental change in the process. In an effort to mitigate the resistance to change when the improvement is implemented, it is crucial to identify the stakeholders early on and to develop a communication plan for each of them. Typical stakeholders include managers, people who work in the process under study, upstream and downstream departments, customers, suppliers, and finance. Regular communication can create more buy-in, identify better solutions, and avoid misconceptions and/or misunderstandings.

Process Map

A process map is simply a step-by-step process of what is happening now in the overall business, a specific department, or an existing process. The process map can follow typical flowchart logic or be a list of steps.

CTQ Factors

Customer requirements and expectations are always considered CTQ. However, the Lean PM also lists any factors that are critical to the success or satisfaction in the project.

CTQ Tree

The purpose of CTQ trees is to convert customer needs/wants to measurable requirements for the business to implement. For example, a retail merchant was receiving a significant number of complaints regarding the homeowner warranty policies from customers. By analyzing customer survey data and developing the CTQ tree, the business was able to identify critical-to-satisfaction requirements. The requirements became the focus for improving customer satisfaction. The business eliminated mandatory warranty visits and made all warranty visits optional. Eliminating mandatory visits satisfied the customers who thought there were too many visits, and adding an extra optional visit satisfied any customers who felt there were too few visits. Expanding the time frame for scheduling warranty visits from two weeks to three months eliminated the inconvenience for customers who had busy schedules and found the time frame difficult to manage. The business took a general, difficult-to-measure need (to improve homeowner warranty satisfaction) and developed specific, measurable, and actionable requirements.

Suppliers, Input, Process, Output, Customer

A SIPOC is a type of high-level process map that includes suppliers, inputs, process, output, and customers. Quality is judged based on the output of a process. The quality is improved by analyzing inputs and process variables. An example of an SIPOC process map is provided in Figure 4.1.

Voice of the Customer

The VOC is a process used to capture the requirements or feedback from the customer (internal or external) to provide them with best-in-class service or product quality. The process is all about responsiveness and constant innovation to capture the customers' changing requirements over time.

The VOC is the term used to describe the stated and unstated needs or requirements of the customer. The VOC can be captured in a variety of

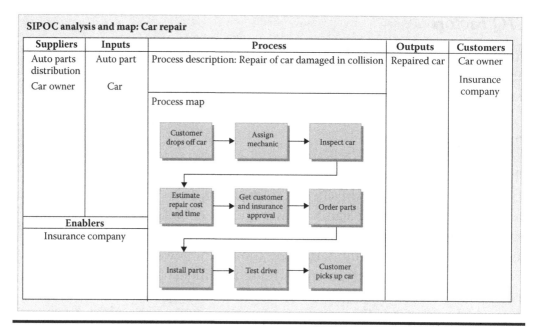

Figure 4.1 SIPOC process map example.

ways, including direct discussion or interviews, surveys, focus groups, customer specifications, observation, warranty data, field reports, and complaint logs. The data are used to identify the quality attributes needed for a supplied component or material to incorporate into the process or product. The VOC is critical for an organization to do the following:

Decide what products and services to offer
Identify critical features and specifications for those products and services
Decide where to focus improvement efforts
Obtain a baseline measure of customer satisfaction against which improvement will be measured
Identify key drivers of customer satisfaction

Following is a list of typical outputs of the VOC process:

A list of customers and customer segments
Identification of relevant reactive and proactive sources of data
Verbal or numerical data that identify customer needs
Define CTQ requirements
Specifications for each CTQ requirement

Affinity Diagram

An affinity diagram (sometimes referred to as a KJ, so named for the initials of the person who created this technique, Kawakita Jiro, Figure 4.2) is a special kind of brainstorming tool. An affinity diagram is used to do the following:

Gather large numbers of ideas, opinions, or issues and group those items that are naturally related

Identify, for each grouping, a single concept that ties the group together

An affinity diagram is especially useful when

Chaos exists

The team is drowning in a large volume of ideas

Breakthrough thinking is required

Broad issues or themes must be identified

SWOT Analysis

An underutilized tool used in Lean as well as Agile is the SWOT Analysis. This is a great project selection tool if there are several projects to consider

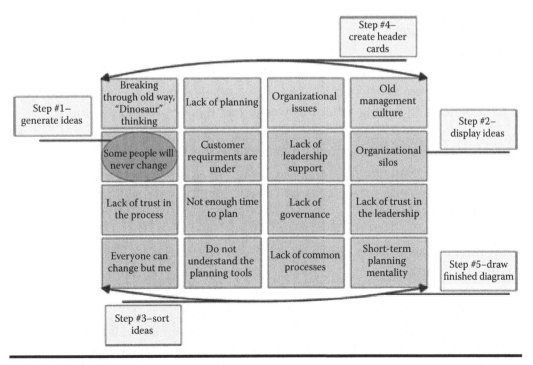

Figure 4.2   Affinity diagram example.

and limited resources. The SWOT analysis emerged in business during the 1960s. It is ambiguous who originally conceived the idea. This tool has become an industry standard and is often used in top management.

SWOT's four components are universally accepted as the following:

Strengths: characteristics of the business or project that give it an advantage over others

Weaknesses: characteristics that place the business or project at a disadvantage relative to others

Opportunities: elements that the project could exploit to its advantage

Threats: elements in the environment that could cause trouble for the business or project

The first step is to introduce the idea or project. In the first quadrant, all the strengths or positives are listed. In the second quadrant, all the weaknesses are listed and so forth (see Figure 4.3).

The SWOT analysis targets the key internal and external factors. Strengths and weaknesses are considered internal to the organization whereas opportunities and threats are considered external factors.

Typical strengths and weakness may include things such as the following:

Human resources

Finances

Internal advantages/disadvantages of the organization

Physical resources

Experiences, including what has worked or has not worked in the past

Typical opportunities and threats may include the following:

Trends

Cultural considerations

Political ideas

Economic issues

Funding sources

Current events

Although the mathematical and benefit prototypes remain valid in project selection, adding facets of Lean and Agile as well as applying LSS tools can drastically expand the way a PM views project selection. There are many

Strengths, Weaknesses, Opportunities, and Threats Analysis

Strengths		Opportunities	
What are your business advantages?	Trusted brand with brand recognition, highly skilled certified tax pros in a professional environment, audit guarantee, options for IRS payments, even advances of the refunds.	Any beneficial trends?	With new tax penalties, more people are looking for a trusted company, people looking for amnesty will need to file returns, community involvement in nonprofits and educational programs, rapid Latino growth to offer ITINs with tax advice, working closely with conciliate and other government programs.
What are your core competencies?	Market leader with available financial and marketing resources, loyal tax pros with long history, loyal customers, dedicated long time tax associates, great leadership with ability to quickly adapt to change.	Niches that competitors are missing?	Name brand, professional upscale office locations, great customer service, free tax reviews, inexpensive audit guarantee, office locations anywhere, online services can be completed in office, professional marketing presence at events.
Where are you making the most money?	Client retention, add on services for revenue, new clients with complex returns.	New technologies?	New software, marketing app available to track event leads, training, better national ad coverage, budget to expand on new technology, customers wanting to use online service using company software but could complete by coming into office to double check return.
What are you doing well?	Revenue is up, 95% brand recognition, high customer service ratings, expanded bilingual offices and Hispanic programs, standardized message on materials which are available to each office, rewarding employees, improved office settings remodeled upscale settings, training associates for taxes.	New needs of customers?	Required to have health insurance, ITINs who pay taxes can file for citizenship.
Weaknesses		**Threats**	
What areas are you avoiding?	Strong presence in the community, marketing activities peak times and pre-season, marketing outside the office, attending community events.	Obstacles to overcome?	Getting new clients to come into offices, local marketing not effective and very inconsistent, distance between offices in rural areas, accountability for activities including tracking results, lack of communication and understanding of marketing goal, change direction quickly based on consumer preferences.
Where do you lack resources?	Time - short run of 12 weeks, available staff for marketing activities with flexible schedule, a method to accurately track weekly local activity results.	Aggressive competitors?	Ability to match deep discounts offered by competitors, people insisting on using a CPA mostly for status, sales training on how to overcome obstacles when talking to clients, new competitors emerging especially small local part time offices.
What are you doing poorly?	Utilizing the short distance between offices to combine marketing resources, communication between offices, competitive pay scale for marketing.	Successful competitors?	More people completing taxes to online which has the threat of identity theft. Also smaller mom/pop tax companies who are open part time but cheat on the return to get more money from IRS.
Where are you losing money?	Bottlenecks at the beginning and end of season, not getting marketing out during the busiest times, unused marketing materials, chamber memberships not used, highly reliant on individual tax pros to bring in new clients with no accountability or measurement.	Negative economic conditions?	More and more people can't afford health care including ACA.
What needs improvement?	Better local marketing activities, accurate updates and reports for weekly activities, Latino growth, clear marketing plan with communication between managers, partnership between offices.	Government regulation?	Health care mandate causing people not to file, confusion and distrust over amnesty programs.
What improvements could help?	Associates that are trained, flexible, branded attire and have the ability to cover 4-5 offices close to each other to standardize the materials and message, staff would need to perform physical labor for event set up/breakdown. Additional staff so that marketing to be done in pairs of two. Must be available and willing to go marketing during peak times when the office is slammed. Incentive or bonus for those choosing to do marketing and finishing the season.	Changing business climate?	Preference to complete taxes online, people less fearful/taking more risks with taxes due to political changes of IRS powers.
		Vulnerabilities?	Unreliable staff (nature of part-time seasonal jobs), a lot of responsibility for one person who would be spread too thin, risk of injury when handling heavy marketing materials, only tax pro can give tax advice.
INTERNAL: What we can control		EXTERNAL: Secondary data	
Operational efficiency, structure, budget, customer service, employee service, capability, resources, talent, process, message		Environment, market trend, tax regulations, changes in laws such as Amnesty program, business, industry data, competitors offers discounts	

Figure 4.3 SWOT analysis.

options and tactics. Remembering, in our newer leaner and more Agile world, that there is no one-size-fits-all solution is vital. Taking advantage of all the tools and thoughts available when making a project selection decision will increase the chances of success.

Most PMs will agree that how a project is initiated determines the success of the project, and yet due to the pressures of working in a reactive society, it is the phase that is economized. Attention to this phase is sometimes more apparent and handled more successfully in Lean and Agile environments as demonstrated above. This is due to the people factor because members of Lean and Agile workgroups are expected to contribute. The ancillary benefit is that buy-in is established from the beginning.

For PMs who do not have the luxury of initiating the projects in the manner discussed earlier in this chapter, a more traditional project management approach will be explored. Whenever possible, tools from Lean and Agile will be applied in an attempt to make the more traditional approach to project management better, faster, and more cost-effective.

In traditional project management, there are essentially five steps in project initiation:

Developing a business case
Undertaking a feasibility study
Establishing a project charter
Forming a project team
Setting up a project management office

Develop a Business Case

There are definitive steps involved in developing a business case. Typically, there are templates available, so recording the information, once it is gathered, is simple. The main items of a business case include the following:

1. Research the business problem or opportunity.
2. Identify the alternative solutions available.
3. Quantify the benefits and costs of each solution.
4. Recommend a preferred solution to your sponsor.
5. Identify any risks and issues with implementation.
6. Present the solution for funding approval.

In Item 1, research the business problem or opportunity, as well as in Item 5, identify any risks and issues with implementation, a simple SWOT analysis discussed earlier in this chapter would accelerate those entries.

Item 2, identify alternative solutions, is best accomplished by first determining what is happening now and performing an analysis of what is working and not working. In Lean, in order to develop a solution, it is necessary to perform an analysis of what is current. The analysis does not have to be complicated. The plan should include the following:

Listing the steps in the process as it currently exists
Mapping the process
Identifying potential causes of the problem
Collecting and analyzing data related to the problem
Identifying root causes of the problem

After data are collected, the original problem statement may need to be modified before moving forward. But the theory behind the bulleted points is that, from this information, a PM should be able to generate potential solutions and, by addressing the root causes of the problem, select a solution.

If the project involves creating something new, the same steps may be applied to another company that is working on something similar. By following the steps and studying the process, Item 3, quantify the benefits and costs of each solution, becomes easy. And finally, Item 6, present the solution for funding approval, is supported by your documented research.

The bottom line of any business case is that CTQs must be considered before submitting the document. CTQ is discussed earlier in this chapter.

Before finalizing this document, it is imperative that the metrics for success are clearly recorded. How will you determine that the project is successful?

Feasibility Study

The next step in initiating a project in a traditional project management environment is conducting a feasibility study.

Is this possible? Will it work? This is the main purpose of conducting a feasibility study. If the PM has truly assessed the current state, a feasibility study may not be necessary. In assessing the current state, sometimes a process in Lean known as the Gemba walk is utilized. This is growing in popularity and can further aid in developing a feasibility study by providing a history of facts. The Japanese term Gemba walk is used to describe personal observation of work where the work is happening.

The approach used for a feasibility study by traditional PMs is, again, form- or template-based and usually uses the following procedure:

Step 1: Conduct a preliminary analysis.
Step 2: Prepare a projected income statement.
Step 3: Conduct a market survey.
Step 4: Plan business organization and operations.
Step 5: Prepare an opening day balance sheet.
Step 6: Review and analyze all data.
Step 7: Make a go or no-go decision.

In each step of the feasibility study, there is a series of prescribed activities as well. For example, in Step 1, the following tasks must occur:

Describe or outline as specifically as possible the planned services, target markets, and unique characteristics of the services
Determine whether there are any insurmountable obstacles

Naturally, there are even more subcategories under these bullets. Satisfying all seven steps in the process can be a lengthy procedure. This may be necessary on a large-scale project, but both Lean and Agile suggest that feasibility on a smaller endeavor might be accomplished just as well by using simple tools, such as surveys and personal or group observation.

In the Lean and Agile environment, PMs would apply VOC as discussed earlier in this chapter.

Project Charter

This is one area in which Lean and traditional project management intersect. Because Agile sometimes handles project selection by group consensus, it does not add much help here. Lean does favor a one-page form, such as a charter, to some of the lengthier charters supported by traditional project management.

Some charters are simply a statement of work. A variety of forms are available, and these templates can guide the PM through the process. Often, it depends on the particular company as to which one is preferred. There are both similarities and differences among these documents.

PMs are often confused by the differences among a business case, statement of work, and project charter. Just remember, in some cases the three documents are combined, and as noted, a variety of templates are available. One, but not always all, of the documents are important in the initiation process.

Setting Up a Project Management Office

In traditional project management, the project management office (PMO) is a department within a business that defines and maintains project management standards for project management within the organization. In many companies, there is not a formal PMO. Throughout the industry, the PMO may not have the same goals and objectives.

In traditional project management education, the role of the PMO serves the following initiatives: (1) identify and solve problems, (2) provide ongoing services to ensure that problems stay solved, (3) cost reduction, (4) provide an efficient centralized service, and (5) standardize materials.

Many Lean and Agile environments do not have a PMO. However, those who do, favor an adaptive approach. An adaptive approach means that the PMO becomes more of an advisory role that offers consulting.

The general role of a PMO, regardless if it follows traditional guidelines or the more adaptive approach favored by Lean and Agile, is to align the selection and execution of projects and programs within the organization's business goals.

Establishing a project initiation strategy is a significant and repeatable activity for the PM. Taking time to make sure that everyone is on the same page and moving in the right direction, when it comes to project selection, will make things easier moving forward. Many Lean and Agile tactics can be adopted to get a more thorough understanding and offer a better roadmap.

Once everything is accomplished, regardless if the PM took a traditional path to project initiation adding Lean and Agile or if the PM originated the initiation by using Lean thinking and Agile techniques, it is necessary to review the results. Before moving to the planning stage, the PM embracing Lean and Agile needs to take time to mistake-proof and reflect on results.

Forming the Team

Sometimes the team is appointed, and sometimes you are the team. As noted earlier, sometimes a PM inherits a team or has to form a team from shared resources. Sometimes the PM does not have the authority to appoint a team nor the charisma needed to inspire team members to join.

In traditional project management, often there is an evaluation process. This process involves interviewing and testing before the final decision is made. Still, even in very structured environments, those selected to be team members for a particular project may have more to do with availability than any other factor. Traditional PMs have a high respect for subject matter experts (SMEs), but the SME normally functions in an advisory role and is not part of the actual project team.

In Lean, more attention is paid to competencies and cross-functionality although traditional project management would certainly support this line of thought. Because Lean concentrates on reducing waste, more consideration is also paid to who needs to be on the team, really. If a team member represents any type of redundancy, then this will eliminate that person. Lean teams often draw from broad-based contributors who may be more generalized than specialized.

Agile teams are cross-functional groups. In many cases, Agile teams have worked on a number of projects together and are not necessarily assigned to a shared resource pool. Subject matter experts are encouraged to be team participants.

Training the Team

One thing that should be considered in this phase and that spills into the next phase—the planning phase—is the topic of training the team. A PM may have a fully trained team, but in cases in which they do not and/or the employee is new to the company, training or bringing the employee up to speed has to be considered. In some cases, the training may simply involve informing the employee of certain criteria or providing a policy manual or a glossary of terms.

Certainly, when a PM decides what training and employee development opportunities should be considered, it is important to first consider the length of the project and if the employees are projected to work on future projects for the company. Training, in general, increases workforce engagement.

Taking a structured project management approach to training has become popular in the human resources field. New concepts in how training should be deployed, delivered, and tracked will come naturally to PMs. This structured method of planning and executing allows trainers to develop a comprehensive process that optimizes resources, people, time, and financial controls.

It is important to remember that even a short-term project may need to be exposed to some type of training. For example, some contracts require special training on hazardous material, and some contracts may want employees to attend an ethics or cultural seminar. In some professions and industries, ongoing training and education is a mandatory requirement.

In all environments, training and education is necessary to improve work performance. Training can be either formal or informal and can take on a blended approach—a mixture of classroom lecture, e-learning, reading, and job shadowing.

Training simulations have become increasingly popular since the methodology and computerized tools to do simulations have become easier to understand and more cost-efficient. Measuring the cause and effect of training is complicated based on what equation you use to get results.

Much has been written about return on investment (ROI) regarding training. In the new economy, the ROI of ROI is being considered as well. The PM must decide how much time should be spent doing ROI exercises. To measure training, it is often satisfactory to put a small amount of metrics in place. ROI on short-format, lunch-and-learn, and demo sessions are generally too time-consuming. Programs that should be viewed through an ROI lens include programs

With high visibility
With hard dollar investments
Regarding compliance issues, such as Sarbanes–Oxley initiatives or ISO certification

There are many different views on employee training. Employers often view this area as an expense that must be tolerated. Some companies only invest training dollars in their high performers whereas other companies choose to support entry-level employees. There are companies who judge their training success on the number of enrollments, and other companies seek a more definitive ROI. So corporate responsibility plays an important role when the PM decides what, if any, training is required.

Although it is highly unlikely that someone will admit he or she does not support training and employee development in the workplace, many managers have reservations. Some believe that it is not cost-effective. Others believe that highly trained employees will seek positions outside the company. There are managers who feel that it is up to the subordinate to seek out training opportunities and that it should not be the responsibility of the company. The PM must understand the company's philosophy on training members of the team.

A well-designed training program pays for itself and increases the bottom line. It maximizes productivity and profits and decreases downtime, equipment damage, and personal injuries. Some side benefits of a strong training policy may also include increased staff loyalty, morale, and motivation. Employees frequently develop a greater sense of self-worth, dignity, and well-being. If the project that the PM is managing is likely to be repeated for another client, the ROI on training will be easier to argue.

Even the most intelligent and skilled employee can benefit a company by participating in safety training, learning better communication skills, and having a stronger understanding of the business.

When initiating a project, Lean and Agile not only looks at company fit and profit but considers the project team, schedule, and education before moving to the planning phase.

In closing, project initiation is the creation of project by the project management that entails the definition of the project's purpose, primary and secondary goals, time frame, and timeline of when goals are expected to be met. The project management may add additional items to the project during the Project Initiation phase.

One aspect that both Lean and Agile appreciate, a little more when initiating a project, than traditional project management, is the consideration of all of those the project will impact. How a project is initiated creates the entire framework moving forward as well as setting the tone for buy-in.

Lean does this by the use of various tools which later can be incorporated in the metrics for success. Agile accomplishes this through face-to-face meetings with the team and listening to various ideas.

Chapter 5

The Planning Process

"Cheshire Puss ... Would you tell me, please, which way I ought to go from here?"

"That depends a good deal on where you want to get to," said the Cat.

"I don't much care where—" said Alice.

"Then it doesn't matter which way you go," said the Cat.

"—so long as I get SOMEWHERE," Alice added as an explanation.

"Oh, you're sure to do that," said the Cat, "if you only walk long enough."

Lewis Carroll, *Alice's Adventures in Wonderland*

The planning phase is the second phase of the Project Management Life Cycle (PMLC). The project manager (PM) enters this phase with several pieces of information. The most important is the project charter (PC). The PC should contain summary aspects of any other documents created and reviewed in the initiation phase. The PC provides the intent of the overall project but not necessarily the direction.

The manner in which a project plan (plan) is constructed depends a great deal on the psychology of the PM and his or her belief about project management theory. Traditional project management (traditional) supports a governed approach that is top down in which decisions are made at the leadership level, and not all knowledge is shared with team members.

It is a plan-the-work and then work-the-plan model, which does not allow for many detours. Agile techniques (Agile) and, in some cases, Lean, on the other hand, promote the following:

Transparency
Frequent inspection
Adaptation

In business models favoring transparency, everyone knows how the project is progressing. The team members who touch the product or service evaluate the process frequently and, based on the findings, collaborate on adjustments. In Agile, adjustments are made quickly.

As with many areas of project management, Lean thinking (Lean) embraces aspects of both traditional and Agile and makes decisions based on the project itself as well as the culture. It has often been remarked that Lean is actually more agile than Agile.

The purpose of a charter for a PM is to get approval, funding, and resources for the project the PM wants to accomplish. The charter is also the first step in building the plan. In traditional and Lean a PC (charter) is a document that outlines the proposed activity in an organized and controlled style. This charter is sometime linear enough that tasks can be outlined immediately in a linear fashion.

Agile promotes people involvement from the beginning and often starts the charter process with the following three concepts:

Vision: The vision defines the "why" of the project. This is the higher purpose or the reason for the project's existence.
Mission: This is the "what" of the project, and it states what will be done in the project to achieve its higher purpose.
Success criteria: The success criteria are management tests that describe effects outside of the solution itself.

When possible, the plan should be attached to the charter to allow the PM to quickly move forward without waiting for separate approvals. Generally speaking, the plan should be recorded using a work breakdown structure (WBS).

Work Breakdown Structure

Most traditional PMs will agree that using a WBS as a way of physically building a project plan is the best approach. A WBS is a key project deliverable that organizes the team's work into manageable sections.

There are many templates available to create a WBS, but it is basically a simple outline. WBSs are also used in education and to express ideas. For example, a textbook may show a WBS as a way to demonstrate what is covered in the material (see Figure 5.1).

A WBS is also a very fast way to develop a draft of a project plan. It is basically a to-do list that can be used to categorize all the things that need to be accomplished under a particular bullet item. By using the format noted in Figure 5.1, it is easy to group like tasks together.

Using the WBS helps the PM to cost out and estimate time for each task. This is helpful when creating a cost or time baseline. The WBS breaks down tasks in a way that resources can be allocated to the specific task with great efficiency.

The Project Plan

In traditional project management (traditional), a worthy plan effectively balances the components of time, cost, scope, quality, and expectations. A plan should incorporate the overall expectations, definition, schedule, and

Basic Project Management Textbook

1 History
2 Approaches
  2.1 The traditional approach
  2.2 PRINCE2
  2.3 Critical chain project management
  2.4 Process-based management
  2.5 Lean project management
  2.6 Extreme project management
  2.7 Benefits realization management
3 Processes
  3.1 Initiating
  3.2 Planning
  3.3 Executing
  3.4 Monitoring and controlling
  3.5 Closing

Figure 5.1   A WBS example.

risks of the project to the organization as well as the blueprint (list of activities). Most of the philosophy and spirit of the plan should be captured in the charter. The charter influences the plan, and the plan records the specific activities along with their associated schedule, costs, and resource needs.

The plan not only shows the activities but indicates how the activities will be controlled throughout the project and any dependencies. Dependencies are tasks that may need to be accomplished prior to a key activity. A PM always wants to stay aware of these situations.

Designing a project plan requires listing all the steps in the process necessary for success. Each step is then assigned a resource, a timeline for completion, and a basic cost. Once the project plan has been reviewed, a time and cost baseline are made. This baseline is used from the beginning to the end of a project to determine if the project is within the acceptable parameters.

Project plans also need to be concerned with constraints (things that could get in the way of project completion) as well as assumptions (things that are assumed will be in place).

Lean supports traditional concepts of charter and plan development. However, in Lean environments, there is a tendency to consider social factors when building a plan.

Whereas Agile techniques take a different approach to the charter and plan development, they do support the Lean mindset that more buy-in is achieved along with more motivated employees if the PM considers the culture and values of the company.

Core values are described as the essential and enduring tenets of an organization. These are the guiding principles that impact how the organization thinks and acts. Company ideology provides the glue that holds an organization together through time.

eBay, the popular online auction house, has five basic values that it posts on its website:

We believe people are basically good.
We believe everyone has something to contribute.
We believe that an honest, open environment can bring out the best in people.
We recognize and respect everyone as a unique individual.
We encourage managers to treat others the way that managers want to be treated.

Values should be universal in the company and govern relationships with employees, investors, customers, and dealers. It is not enough to determine values. The entity must be willing to invest in resources and tools necessary. They must be willing to reward behavior that supports these values.

Harley-Davidson, the highly respected motorcycle manufacturer, espouses five company-wide written values simply stated as the following:

Tell the truth
Be fair
Keep promises
Respect the individual
Encourage intellectual curiosity

But Harley-Davidson does not stop there. It lists six behaviors that support an ethical decision-making process:

Allocate resources to learning
Encourage risk taking
Challenge the status quo
Benchmark performance against the best in class
Be open to influence
Accept responsibility for lifelong learning

Many companies list their values as something that sets them apart, making them unique and special. Some companies print and post their values. Most business values focus on the customer or personal integrity.

In Lean thinking and Agile technique (Lean and Agile) environments, a PM will often function in many roles, sometimes participating hands-on in a project and not just as manager. Therefore, a PM has an added responsibility of leading the group when it comes to supporting the company values. Even when there is no ethical breach, a PM has to be careful not to allow the perception that behavior is unethical or not in line with the company's core values.

Because most companies make decisions according to a few core values, a PM must also be aware of these values because values help people embrace positive change.

As noted earlier in this chapter, Agile handles the charter and plan process differently than traditional or Lean.

Lean and Agile agree that placing an emphasis on the people factor is important in the charter process. Whereas Agile may have a less formal charter focused on ideas and thoughts, the charter in both the Lean and Agile world dictates the project plan.

Traditional project management as well as Lean place a strong emphasis on the WBS when developing a plan. Agile favors more ongoing conversation and communication as opposed to a specific defined plan and supports this with more graphical representations. This is not to say that Agile does not use a WBS as a method. In fact, in Agile groups, Gantt charts are very prevalent. A Gantt chart is a visual representation of a WBS.

For the traditional PM wanting to embrace a few Lean or Agile tools, the most important thing to remember is the work being performed is the same. All PM focus should be on what needs to be accomplished for a project to be on time and on budget. This thinking does not change with different project management theories or models.

In traditional, and sometimes in Lean, the phases of a traditional Waterfall development force a PM to the next phase only when the previous one is complete. The plan reflects this.

In Agile, project development is an ongoing process of making decisions based on the realities observed in the actual project. This can be beneficial when a plan can be flexible. Agile is not always the best decision for a plan that contains a lot of compliance factors. This is because documenting and adhering to the process in compliance-related issues is often more revered than the actual outcome.

Traditional, Lean, and Agile can be blended to get the best charter and plan to achieve the ultimate goal, which is value to the customer.

The project often starts with a phrase called the problem statement. In many cases, it is a problem that needs to be solved. However, the problem statement can also be a declaration of what needs to be created or improved. Therefore, a problem statement might not always meet the criteria of what is considered a problem. The resolution, modification, or creation of a new process is referred to as the solution. A good PC has a well-defined problem statement as well as a solution.

However, if the PM is not provided with a solution, there are a number of problem-solving methodologies available. Lean Six Sigma promotes the Define, Measure, Analyze, Improve, Control (DMAIC) model discussed in Section III whereas Agile supports a more critical to quality (CTQ) and voice of the customer (VOC) approach. Another problem-solving method involves simulation, which may or may not be computer-generated. This is where a solution is hypothetically applied, and the outcome is measured with theoretical data.

Using prototypes as a framework to brainstorm planning is not only an effective way to solve a problem, but it also helps provide the logic when explaining why a particular course of action was chosen. Many of the models below overlap and/or use mutual tools. And some models highlighted in this chapter are identified as being more common in a certain discipline, for example, traditional project management (traditional), Lean thinking (Lean), or Agile techniques (Agile).

A thought to consider is that traditional favors more structured approaches whereas Agile promotes creative, flexible, people-rather-than-process approaches. Lean methods tend to fall somewhere in the middle.

Models for Planning

The following models can be used in either initiating or planning. They are particularly useful when the PC does not provide enough direction to use a WBS to determine the tasks that need to be accomplished.

Each of these popular prototypes also notes the type of project management most often associated with that model: traditional, Lean, or Agile. Many models are similar, and some overlap:

Structured problem solving traditional
Plan-do-check-act (PDCA) model: traditional, Lean
DMAIC: Lean Six Sigma
People-process-product/service method (3Ps): Lean
What-why-where-who-when-which (6Ws) approach: Lean
Machines-methods-materials-measurements-mother nature-manpower
   (6Ms) method: Lean
A3 format: Lean and Agile
5 Why approach: Lean and Agile
Six Thinking Hats®: Agile

Structured Problem Solving When Planning

Structured problem solving is popular and embraced by many business management theories and is not limited to traditional. Although tools may vary slightly, the steps include the following:

1. Decide on which problem to pursue.
2. Define the problem.
3. Determine the root cause.
4. Generate possible solutions and choose the most likely one.
5. Plan and execute the solution.
6. Verify effectiveness.
7. Communicate and congratulate.

When using this model as a hypothetical, Steps 1–4 are still performed as actual activities. But Steps 5–7 are used for discussion-based what-if scenarios.

PDCA Model

In traditional project management as well as Lean, another popular way to develop a solution is to run the problem through a simulated PDCA model (see Figure 5.2).

The PDCA cycle is made up of four steps for improvement or change:

1. Plan: Recognize an opportunity and plan the change.
2. Do: Test the change.
3. Check: Review the test, analyze the results, and identify key learning points.
4. Act: Act based on what is learned in the check step.

This model is known by several different names, which include the Deming wheel and the Shewhart cycle. Both Edwards Deming and Walter Shewhart promoted the model and are considered pioneers in the quality movement.

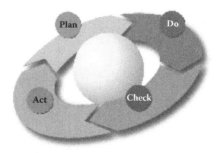

Figure 5.2   PDCA model.

The advantage of using PDCA for problem resolution is that the model is simple. It does not take a lot of discussion or training for a team to understand the configuration. This model is also helpful for conducting a pilot during the testing phase.

DMAIC Model

DMAIC is an acronym for five interconnected phases: Define, Measure, Analyze, Improve, and Control. It is used in Lean when Lean is combined with Six Sigma. This is an overview. In Section III of this book, the DMAIC processes and tools are discussed in depth.

Six Sigma business philosophy employs a client-centric, fact-based approach to reducing variation in order to dramatically improve quality by eliminating defects and, as a result, reducing cost. Because the DMAIC is a more complex problem-solving model, more detail is required. Here is a brief description of what takes place in each phase.

The Define phase is when a team and its sponsors reach agreement on what the project is and what it should accomplish. The outcome is the following:

A clear statement of the intended improvement
A high-level map of the processes
A list of what is important to the customer

The tools commonly used in this phase include the following:

Project charter
Stakeholder analysis
Suppliers, input, process, output, and customers (SIPOC) process map
VOC
Affinity diagram

The Measure phase builds factual understanding of existing process conditions. The outcome is the following:

A good understanding of where the process is today and where it needs
    to be in the future
A solid data collection plan
An idea of how data will be verified

The tools commonly used in this phase include the following:

Prioritization matrix
Process cycle efficiency
Time value analysis
Pareto charts
Control charts
Run charts
Failure mode and effects analysis (FMEA)

The Analyze phase develops theories of root causes, confirms the theories with data, and identifies the root cause(s) of the problem. The outcome of this phase includes the following:

Data and process analysis
Root cause analysis
Being able to quantify the gap opportunity

The tools commonly used in this phase include the following:

5 whys analysis
Brainstorming
Cause and effect diagram
Affinity diagrams
Control charts
Flow diagram
Pareto charts
Scatter plots

The main purpose of the Improve phase is to demonstrate, with fact and data, that the solutions solve the problem.
  The tools commonly used in this phase include the following:

Brainstorming
Flowcharting
FMEA
Stakeholder analysis
5S method

The Control phase is designed to ensure that the problem does not reoccur and that the new processes can be further improved over time.

The tools commonly used in this phase include the following:

Control charts
Flow diagrams
Charts to compare before and after such as Pareto charts
Standardization

The Control process involves quality and statistical concepts that have existed for decades. However, the advent of quality control software makes the process simple enough for anyone to perform.

Variation is everywhere, and it degrades consistent, good performance. Valid measurements and data are required foundations for consistent, breakthrough improvement.

Having a standard improvement model, such as DMAIC, provides teams with a roadmap. The DMAIC is a structured, disciplined, rigorous approach to process improvement, consisting of the five phases mentioned, and each phase is linked logically to the previous phase as well as to the next phase.

When using the DMAIC (Figure 5.3) as a problem-solving model, the PM should focus on the Define, Measure, and Analyze phases. At the end of the Analyze phase, the DMAIC model yields three to five solutions. In the Improve phase, a solution is chosen and piloted.

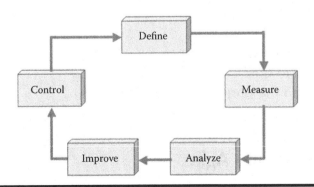

Figure 5.3   DMAIC model.

3Ps Method

The 3Ps method considers three categories: people, process, and product/service. This approach is often used in Lean and captured in a fishbone diagram. It is also used as a way to facilitate seeing the problem through three perspectives. The product/service lens is just a convenience to help the PM remember that the third perspective lens, or bone on the fishbone diagram, could be product or service.

6Ms Method

In the 6Ms model, machines equal all hardware. Methods are equivalent to any policies or procedures (real or proposed). Materials account for all nonhuman resources. Measurements represent any measurements or measurement analysis. Mother nature signifies environment and environmental conditions. The term manpower refers to the people factor.

The 6Ms method is similar to the 3Ps. However, the 6Ms is a more detailed approach. As with the 3Ps, it makes a good fishbone diagram, but seeing the problem through different viewpoints is the true value.

6Ws Approach

The 6Ws approach uses the questions adopted by journalists. When a PM needs to publish a report or make a presentation, considering the 6Ws can help create focus and organization. This tool, used mostly in Lean, can also facilitate conversations and brainstorming activities with team members.

What: What will you make?
Why: Why will we do this?
Where: Where will this happen?
Who: Who will do this?
When: When will the project start and stop?
Which: Which approach will you take?

A3 Format: Lean and Agile

The A3 format refers to a process originally developed by Toyota as a template. The template was used for three different types of reports: proposals, status, and problem solving.

Over the years, the A3 format has been adopted in both Lean and Agile environments to solve problems. A3 formats differ slightly, but all are based on the PDCA model and generally include these steps:

1. Identify the problem or need.
2. Understand the current situation/state.
3. Develop the goal statement—develop the target state.
4. Perform root cause analysis.
5. Brainstorm/determine countermeasures.
6. Create a countermeasures implementation plan.
7. Check results—confirm the effect.
8. Update standard work.

Most of the steps in this process are intuitive, and a PM can immediately identify what needs to be done. The term standard work is a Lean term to describe the detailed definition of the most efficient method to produce a product or service. The factors considered in designing standard work include documentation that is current, complete, clear, correct, and concise. As a better practice is developed, that becomes the standard work.

5 Why Approach: Lean and Agile

Originally used in Lean, this approach has been adopted by Agile as well as many other problem-solving methodologies. Most think of it as a tool. However, the 5 Why approach can uncover the root cause of a problem, so effectively, a solution is evident, earning it the right to be considered a problem-solving model.

In the 5 Why approach, a person is asked the question "why?" five times. Each answer that is given is incorporated into the next question. For example, if the problem being tackled involved equipment being shipped to the wrong location, employees in the shipping department would be queried by the supervisor: "Why is equipment being shipped to the wrong location?"

In this scenario, perhaps a shipping employee responds:

"Because we do not have the correct labels."

Then, the next question asked by the supervisor would be "Why do you not have the correct labels?"

The shipping employee answers, "Because the managers do not fill the labels out correctly."

The supervisor might question (the fourth why) "Why do the managers not fill out the labels correctly?"

The answer to this question may prompt the final, or fifth, why, or it may suitably answer the question in a way that the solution is obvious.

But if the next answer by the shipping employee is "because the codes changed and managers were not alerted," the solution might be to create forms that show the codes until managers become familiar with the new ones.

Or the solution might be training. Or the solution might be that shipping employees fill out the labels. The point is that the answer to the fifth why may lead to several solutions.

Six Thinking Hats

Although most Lean practitioners have been exposed to the Six Thinking Hats, the theory is more popular among supporters of Agile. In Edward de Bono's book explaining problem solving, he believes, similar to 3Ps and 6Ms, that viewing a problem from various standpoints expedites the answer. These angles take a rather humorous approach and are related to various colors. They include the following:

Information (white hat): Considering purely what information is available, what are the facts?

Emotions (red hat): Intuitive or instinctive gut reactions or statements of emotional feeling.

Bad points judgment (black hat): Logic applied to identify flaws or barriers.

Good points judgment (yellow hat): Logic applied to identifying benefits, seeking harmony.

Creativity (green hat): Statements of provocation and investigation, seeing where a thought goes.

Thinking (blue hat): Used to manage the thinking process, a control in place to ensure guidelines are observed.

There are several ways that a PM can develop a solution. A PM may need to use a more formalized approach, such as structured problem solving, PDCA, or DMAIC. For a compliance or highly complicated project, the PM may want to consider the DMAIC because of the following benefits:

Better safety performance
Effective supply chain management
Better knowledge of competition and competitors
Use of standard operating procedures
Better decision making
Improved project management skills
Sustained improvements
Alignment with strategy, vision, and values
Increased margins
Greater market share
Fewer customer complaints

There are also benefits to using Lean or Agile models. The 3Ps, 6Ms, and 6Ws are all useful when facilitating information-gathering sessions and capitalizing on collective intelligence. A3 formats help with reasoning. The Six Thinking Hats help with perceptions and outlooks.

A PM will want to consider a variety of models to solve a problem if a solution is not apparent. If the PM knows what needs to be done, the next step is to develop the project plan.

If the PM is unsure how to solve the problem or has many options, considering a model driven by the outcome, environment, history, and culture is advantageous. Understanding various models provides the PM with valuable options.

In the Lean and Agile environment, the PM must also be concerned with the organization. Many decisions on how much attention should be directed to this area will once again depend on the length of the project and how the project fits into the overall schema of the company.

Certainly, training is a component that may have already been addressed in the initiation phase. But organizational fit and effectiveness need to be considerations when planning a project. This would include any of the cultural issues that may become obstacles. The Lean and Agile PM knows that in order to have a successful project these factors should be addressed up front.

The quickest way the PM can assess the situation is to ensure that the project is in alignment with the company goals. Because the PM may have limited time to spend in this area, a model often applied to assessing training needs is a good place to get started. This is a quick way to assess organizational development (OD) and project needs. When time is short, an instructional systems design model that can be applied to OD is ADDIE. This model is similar to that used in the Project Management Body of Knowledge™ (PM-BOK). ADDIE simply means the following:

Analysis
Design
Development
Implementation
Evaluation

Here is a brief summary of the process. Keep in mind that the Lean and Agile PM may simply use this as a critical thinking tool to reason through the best approach to quickly train a staff member. This instructional systems design model can also be used for more complex or long-term training initiatives.

Analysis

The analysis phase clarifies the learning opportunity and identifies the learning environment and learner's existing knowledge and skills. Questions the analysis phase addresses include the following:

Who are the learners, and what are their characteristics?
What is the desired new behavior?
What types of learning constraints exist?
What are the delivery options?

Design

Design deals with learning objectives, assessment instruments, exercises, content, subject matter analysis, lesson planning, and media selection. It is systematic and specific. Systematic means a logical, orderly method of identifying, developing, and evaluating a set of planned strategies targeted for attaining the project's goals. Specific means each element of the instructional design plan must be executed with attention to details.

Development

In the development phase, instructional designers and developers create and assemble content assets blueprinted in the design phase. In this phase, the designers create storyboards and graphics. If e-learning is involved, programmers develop or integrate technologies.

Implementation

The implementation phase develops procedures for training facilitators and learners. Trained facilitators cover the course curriculum, learning outcomes, method of delivery, and testing procedures. Preparation for learners includes training them on new tools (software or hardware) and student registration. Implementation includes evaluation of the design.

Evaluation

The evaluation phase consists of two aspects: formative and summative. Formative evaluation is present in each stage of the ADDIE process, and summative evaluation is conducted on finished instructional programs or products.

The topic of training and organizational fit and effectives can be addressed in the initiation or planning phases of the PMLC. The Lean and Agile PM knows that, in addition to all the technical requirements necessary to plan a project, attention and detail need to be given to the people factors and organizational considerations involved.

Chapter 6

Project Execution

"My dear, here we must run as fast as we can, just to stay in place.
And if you wish to go anywhere you must run twice as fast as
that."

The Queen of Hearts, in Lewis Carroll,
Alice's Adventures in Wonderland

Project Managers (PMs) generally agree that a project plan should be piloted
or tested prior to execution. Traditional project management (traditional) and
Lean thinking (Lean) have formal attitudes toward pilots and testing. Agile
techniques (Agile) have a different spin on how pilots and testing should be
conducted. Some project management life cycle (PMLC) approaches include
the following information in the planning phase of the project. Regardless,
before a project is rolled out to the enterprise, it should be tested or piloted
in some way to ensure success. This concept is strongly emphasized in Lean.

In Agile, piloting and testing are ongoing. A great deal of attention is paid
to the voice of the customer (VOC). Agile has the ability to adapt a project
plan on the fly to meet or exceed a customer expectation.

Whereas this chapter will focus mostly on traditional and Lean methods
and how they can work together or enhance the piloting and testing agenda,
Agile testing is briefly covered next.

Performance testing in Agile does not align with traditional or Lean but
offers many thoughtful processes that traditional or Lean PMs may want to
incorporate. One such measure is quadrant testing. The Agile testing quadrants (the quadrants) are based on a matrix Brian Marick developed in 2003
to describe types of tests used in extreme programming (XP) projects.

The quadrants are simply to make it easier to collect information and do not follow a Waterfall path. Quadrants are divided into these four categories:

Q1: Technology-facing tests that guide development
Q2: Business-facing tests that guide development
Q3: Business-facing tests that critique (evaluate) the product
Q4: Technology-facing tests that critique (evaluate) the product

At this juncture, the quadrants are primarily used for technical projects. The quadrants then consider themes. To understand the concept of themes, there are two other terms to consider: user stories and epics.

The user story describes the type of user, what they want, and why. A user story helps to create a simplified description of a requirement. An epic captures a large body of work. It is essentially a large user story that can be broken down into a number of smaller stories. A theme, which is what testing quadrants are primarily addressing, is a group of user stories that share a common attribute, and for convenience, they are grouped together.

The testing quadrant in Agile is considered a standard way to test and track performance themes but is not embraced by all Agile practitioners.

The purpose of a project pilot or a test prior to rolling out the project is to identify and manage risk, validate the benefit, and secure buy-in. Whereas all project management theories support the idea of piloting or testing, as noted earlier, traditional and Lean take a formal approach. But it appears that all project management works place more emphasis on evaluating the pilot or test than on the piloting or testing strategy. Traditional supports the following steps:

State the goal based on the business purpose that is driving the project.
Develop an implementation plan and solution scenarios and build prototypes.
Develop an evaluation metric and decide what measurements need to be accomplished for piloting or testing to be considered successful.
Determine the timeline and cost baseline for the activity if the test will take more than 30 hours.

There are also a number of administrative tasks that need to be considered when conducting a test. These tasks are related to things such as staffing, schedules, and facilities. Deciding on which resources will participate in the piloting or testing depends on the size and scope of the test. In large-scale piloting or testing, human resource selection may need to depend on statistical sampling. In order to use sampling, a clear definition of the testing purpose has to be established and documented.

Sampling is the process of selecting units (e.g., people, organizations) from a population of interest. Studying the sample responses provides feedback as to the potential success or failure of a project or idea.

There is a type of pilot that has some variations from typical pilots and is used by traditional and Lean. This is known as proof of concept (POC).

A POC is designed to demonstrate the feasibility of a proposed idea or concept to solve a business need. POC may also be used during the project selection process. The disadvantage to using POC is that generally it is so specific to the particular product or service that in order to conduct a valid study, subject matter experts in that topic need to be available. Also, there is a great deal of variation when conducting a POC. Because the purpose of POC is to prove that a project or idea will work, several one-off strategies specific to that particular product or service may be employed.

However, the basic steps for a POC are amazingly similar to those of a regular pilot, but they tend to contain much more detail. Typically, a POC scheme would include the following:

Step 1: Define and develop
  Identifying stakeholders and team members
  Defining goals, inputs, objectives, scope, and success criteria
  Establishing resource commitments and finalizing a POC schedule
  Work with stakeholders to prioritize functionalities
  Determine deliverables
Step 2: Engineer
  Configuring and testing the required infrastructure
  Determining solution steps
Step 3: Execute
  Creating the test design for use cases and defining positive and negative
    test scenarios
  Designing and executing test scripts
Step 4: Evaluate
  Reviewing and validating the POC results with all stakeholders
  Comparing outcomes to success criteria and aligning implementation if
    appropriate

Once a PM has developed a solution to the problem statement, it may be tempting to skip piloting or testing. In projects involving software development, testing is not an option. However, in projects that do not have an information technology (IT) component, it is not uncommon that this activity is avoided.

There are a number of reasons for dodging the endeavor, but most often, the reasons are time, money, and scope. These are referred to as the triple constraints. The triple constraints are real concerns, but bypassing piloting or testing will only enhance these constraints and may lead to additional obstacles. Industry definitions explain time, money, and scope as follows:

Time: This refers to the actual time required to produce a deliverable.
Cost: This is the estimation of the amount of money that will be required to complete the project. Cost encompasses various things, such as the following:

Resources
Labor rates
Risk
Materials
Scope

These are the functional elements that, when completed, make up the end deliverables for the project. Lean and Agile both offer ideas on piloting or testing schemes. For example, some of the simulations discussed in the previous chapter on developing a solution can be modified to conduct pilot testing.

In addition, other Lean thinking and Agile techniques (Lean and Agile) methods can be adapted to test a project's potential success and identify risk.

Before a project plan is executed, every attempt should be made to mistake-proof. Mistake-proofing is a Lean concept. In Japanese, it is known poka-yoke. Poka-yoke means the use of automatic devices or methods that either make it impossible for an error to occur or make the error immediately obvious. A common example of poka-yoke is modern electronics in which connects are color-coded to the plugs.

The following mistake-proofing/poka-yoke steps have been modified to accommodate a project plan.

Create a high-level flowchart of the major steps in the proposed project. Review each step, thinking about where and when human errors are likely to occur.

For each potential problem, work back through the process to find its source and think of things that can be written into the project plan that will minimize that risk.

If possible, eliminate the step that is the root cause of the error.

Make the corrective action easier than the error.

A closely related concept that is sometimes even included in mistake-proofing strategy and also popular in Lean is the idea of inspection or observation. Inspecting and/or observing things, such as materials needed, human resources involved, and facilities, can also help test a project in realization and identification of threats.

A formal source inspection check before your proposed project activity takes place to check that conditions are capable of handling the task is a quick but powerful way to set controls, especially if controls are automatic and able to keep the process from proceeding until conditions are right.

A less formal way to test a project's success that also helps with communicating the project is thinking in terms of voices. The primary voices are the following:

Voice of the customer (VOC)
Voice of the employee (VOE)
Voice of the process (VOP)
Voice of the business (VOB)

Most PMs are aware of the importance of VOC. The VOE can offer a great deal of insight by simply asking certain employees to review the project plan draft. Often it is the employee who may be able to identify a risk the PM has not considered. The VOE is also useful to identify physical bottlenecks or undocumented processes.

When using VOP, the PM needs to carefully determine if the current process is able to handle the task recorded on the project plan. And finally, VOB, which stands for voice of the business but really means voice of the industry, needs to be considered from a compliance standpoint.

Traditional project management, as noted earlier, offers a number of tips on evaluation metrics.

Evaluation Metrics for Piloting or Testing

Many of the metrics used to measure a pilot or test can also be used to measure the overall success of the project. Performance metrics are the most important ones when assessing a pilot. These include, but are not limited to, the following:

Schedule and effort/cost variance
Resource utilization
Change requests to scope of work
Number of problems reported by attendees

Schedule and Effort/Cost Variance

Schedule and effort/cost variance measure the performance as well as progress of the project against signed baselines. If a pilot will take more than 30 hours to complete, not including the evaluation, a timeline and cost baseline should be established. In traditional project management, a schedule showing effort and cost variance (projected at that point in time and actual at that point in time) is often called earned value. This metric integrates project scope, cost, and schedule measures to help the PM assess and measure project performance and progress. This metric uses past performance (i.e., actuals) to more accurately forecast future performance. There are three main factors to consider:

Planned value (PV)
Earned value (EV)
Actual cost (AC)

PV is how much was planned to spend for the work or is budgeted to accomplish the work breakdown structure. EV is the value of work performed to date. AC is the actual cost to date. Using these three variables, schedule and cost variance can be analyzed.

EV is a method that allows the project manager to measure the amount of work actually performed on a project beyond the basic review of cost and schedule reports.

The formula to calculate the earned value is simple. Multiply the actual percentage of the completed work by the project budget. Earned Value = % of completed work × budget at completion (BAC).

Resource Utilization

The next metric, resource utilization, simply measures the productivity of resources. There are a number of strategies available to do this, and they are transaction-based or rely on observation. Resource utilization in project management focuses on how much time team members spend on various tasks.

The easiest formula is:

1. Resource utilization = busy time/available time.
2. Resource utilization = planned working hours (bookings)/available hours.
3. Resource utilization = recorded working hours/available hours.

Change Requests to Scope of Work

In project management, scope is the extent of work needed to finish a project that specifically states the expectation. Changes in scope can interfere with the time and cost baselines.

Change requests to the scope of work should be very limited in the actual project but may be prevalent in the pilot due to the fact that some pilots are flexible by design, especially the handful conducted in Agile environments. Incorporating this metric can help determine where problems may potentially occur in the project and can also be useful when designing frequently asked questions or training.

Number of Problems Reported by Pilot Attendees

The number of problems reported by pilot attendees is an intuitive measure that should not be discounted. Listening to VOE can save rework.

Agile considers the testing function from a different stance. Agile believes that testing is not a separate phase, but an integral part of the project that sometimes functions in tandem with process steps. Evaluation is ongoing.

Testing in Agile is sometimes performed by checking VOC responses on iterative bases to ensure expectations are being met. Because Agile promotes itself as flexible by checking in on VOC on a regular basis, it is possible to change gears and rewrite plans as needed.

A testing technique that could be considered off-label is using the failure modes effects analysis (FMEA) psychology. This is where the PM would take the core part of an FMEA form and apply it to critical or crucial events in the project. These events would be described as ones that if they did not go smoothly could interfere with the timeline or cost baseline.

To understand the power of the FMEA psychology, it is necessary to understand the overall FMEA function. An FMEA is a step-by-step approach for identifying all possible failures. FMEA forms typically contain the following:

Failure modes (what could go wrong?)
Failure causes (why would the failure happen?)
Failure effects (what would be the consequences of each failure?)

The FMEA form contains a number of checks and balances and often assesses activities within a project or department in the framework of risk, number of occurrences, and detection rate. Detection rate represents how likely a condition is to occur without being noticed.

In the FMEA psychology model, only the bullet points are considered and applied to those tasks that are absolutely necessary for the project's successful completion. Then, based on the discoveries, items may be written into the project plan to either correct immediately or inform the PM with enough notice to correct.

Piloting or testing is essential to ensure a project will be completed on time and within budget. Piloting or testing strategies depend on the sophistication and factors in a project. PMs can learn from both Lean and Agile when conducting this exercise.

It can be uncomfortable to move from piloting or testing to execution. But there are markers that indicate that this activity should be completed. Some are not optional. An example would be a depleted budget or an approaching deadline.

If the PM enters the execution phase with a strong project plan (plan) and a successful pilot, it would be reasonable to assume that things should become easier. After all, the work breakdown structure (WBS), if done correctly, should reflect tasks in sequential order and who is responsible for each task. The PM's main objective in this phase is executing, monitoring, and controlling phases of the PMLC, which consists of completing and managing the work required to meet the project objectives. This phase also ensures that the project performance is monitored and adjustments to the project schedule are made as needed.

However, the role of the PM has changed drastically over the past two decades. For example, historically, a PM sent most employee-related issues to the human resources (HR) department. Executing, monitoring, and controlling the plan requires incredible people skills.

Now, PMs are expected to handle many of the day-to-day issues with their teams. A PM originally had a staff that included a project deputy and/or a project secretary. The positions have evaporated, and a PM, more times than not, is responsible for tracking project movement and doing a host of administrative tasks.

As discussed earlier, a PM is now expected to handle many situations without direction or training related to communication zones. Examples include areas such as the following:

Coaching
Mentoring
Mediation or arbitration
Dispute resolution

A PM may be required to deal with certain aspects of employee development originally handled by a training department or HR. A PM may even be responsible for identifying learning opportunities and formulating training strategies for their subordinates. Today, a PM often absorbs the role of a safety officer and handles more risk management issues than ever before.

Cultural and diversity matters are quickly becoming topics that a PM cannot ignore. The workforce has become older, causing generational differences. The workforce has become more international, requiring extra sensitivity. A PM is often responsible for communicating a project plan (plan) to a group of people who are needed to perform certain project tasks but who do not report directly to a PM. Work in this century has become very individualized. A PM must consider the challenge of managing virtual teams and individuals who telecommute.

PMs must often rely on their personal influence to gain buy-in and productivity rather than relying on HR policies.

Employers who include disability issues in corporate diversity policies enrich and enhance workplace benefits in the new economy. However, PMs now are charged with the responsibility of committing to and implementing these policies. It is up to managers to design disability-friendly strategies for the workplace, which often includes education of the staff.

Being a compliance officer, employee coach, and performance manager has added new stress to the job of a PM, and there are still management duties, such as resource leveling, finance, and marketing. The information provided in the following chapters is designed to create a balance and to provide resource information.

In these cases, the basic communication skills of a PM become more valuable than producing a stellar plan. In fact, rock-solid graphs, charts, and narratives can work against a PM in an environment that does not truly understand project management theory. Pushback from team members who may or may not be part of the project team can view a PM as overcomplicating and making their daily work life more difficult.

Even in Agile technique (Agile) situations in which conversations and observations are favored over documentation and other paperwork, there is some written accountability necessary on any project. Resistance to recording anything or taking on extra work is not uncommon. All projects represent change. Change, as a PM knows, is not always enthusiastically embraced.

Projects have the most likelihood of failing in the execution phase. And, not surprisingly, it is due to people issues. Much of this chapter has to do with communication. The plan, by this phase, has been developed, and now, all that needs to be done is the rollout.

A PM has to be concerned with managing plan performance by communicating the plan to the workgroup. Without determining or explaining this, communicating the plan becomes an announcement that is easily ignored. Performance management is a science that takes the emphasis in organizations away from command and control toward a facilitation model of leadership and is adopted informally both by Lean and Agile.

Execution of the plan involves compensation issues, availability of tools, report structures, and, in some cases, even environmental issues, such as lighting, temperature, and location of the facility.

The overall topic of communication fills volumes. But a good place to start is with the history. The following concepts are summarized and form the foundation for communications studies. Understanding how human communication works, in general, will help in communicating and rolling out the plan.

Stuart Hall, a social theorist, made several observations about communication. Hall's paper "Encoding/Decoding," published in 1973, provided major influences on cultural studies. In human communication, according to the encode/decode model of communication, what makes communication possible is a common language. Hall's model focused on three components: the sender, the channel, and the receiver.

For several years after the paper was published, international cultural training focused on learning language phrases to show you were trying to communicate along with a few basic customs and traditions of that country. This type of training focuses on encoding and decoding.

Eventually, most professionals associated with cultural awareness agreed that the model was not a true representation of how people communicate. One of the main criticisms of encoding and decoding is that it is too linear and does not apply to the international community. In reality, the roles of the sender, channel, and receiver continually alternate. For example, how the human receiver interprets a message and assigns meaning is based on many factors in addition to a common language, such as the following:

Intention
Relationship between sender and receiver
Context of the message

There are additional flaws with the encode/decode model. Serious thought should still be given to the sender, the channel, and the receiver for successful communication. The sender, for example, should be aware of any personal biases that might hinder communication. The channel needs to be appropriate. Some cultures respond to electronic communication whereas others prefer a more personal channel, such as conversation. The receiver's social or political standing in the company might also impact communication. Some cultures move quickly to the point; others talk things through long enough to establish rapport or a relationship with the other person.

Certainly, learning the language helps. Then again, the spoken word only accounts for part of the message conveyed. Other considerations include body language, facial expression, and context of the conversation. Internationally, all of these factors have the possibility of being misinterpreted. Body language is sometimes the most difficult to control. However, it is essential that a PM discover how that particular country views the following:

How far away to stand
How much eye contact is appropriate
What gestures are considered inappropriate
What are appropriate emotional responses

How people are addressed, formally or informally
Table etiquette

In the *Mathematical Theory of Communication*, written for electronic communication, mathematician Claude E. Shannon determined these basic elements of communication:

An information source that produces a message
A transmitter that operates on the message to create a signal that can be sent through a channel
A channel, which is the medium over which the signal carrying the information that comprises the message, is sent
A receiver, which transforms the signal back into the message intended for delivery
A destination, which can be a person or a machine for whom or which the message is intended

This model was designed for electronic systems. But components of this model may be applied to individual communication. The model expands on the encoding/decoding human communication model.

As noted earlier in this chapter, there are many ways to communicate a plan, and most depend on the project management theory adopted by a PM. Traditional often relies on announcements after the fact whereas Lean and Agile start communicating the project parameters early in the process.

One way to communicate data quickly is by designing and then sharing a document collection plan. Usually, there is a data collection period involved in a project during which information gathering is critical to the project's success. Agile is constantly communicating with team members through Scrum and sprint sessions, but even Agile supporters sometimes forget that during the data collection period team members are often concerned about why the activity is taking place and may become unnecessarily suspicious.

Executing the plan requires keeping employees on task to adhere to the plan, but it also involves good communication skills, team motivation, and being flexible.

Once a project moves into the execution phase, the project team and the necessary resources to carry out the project should be in place and ready to perform project activities. The project plan should have been completed and baselined by this time as well. The project team and specifically the PM's focus now shifts from planning the project efforts to participating in, observing, and analyzing the work being done.

Particular attention during project execution will need to be paid to keeping interested parties up to date with project status, dealing with procurement and contract administration issues, helping manage quality control, and monitoring project risk.

The critical project management elements for the project team to help with include the following:

Performance monitoring
Provide project status

Performance Monitoring

This function implements an execution plan to measure actual performance as compared to planned performance. Actual project schedules will need to be reviewed periodically and compared to baseline schedules.

Provide Project Status

The PM is responsible for relaying project status to parties outside the project team; the project team is expected to report status to the PM. This includes communicating information on both a formal and informal basis.

Without a defined project execution process the Lean and Agile project manager cannot ensure that team members are executing projects independently and using their own experience. The plan should reflect how tracking and corrective action activities should be handled.

It is also critical during the execution phase that the PM supports and monitors the execution of other important project plans, such as the communications plan, the risk plan, and procurement plan.

The Lean and Agile project management is better equipped to handle project plan modifications. These changes could result from the following:

New estimates of work still to be done (generated as more detailed information is known about outstanding work)

Changes in scope/functionality of end product(s)
Resource changes
Unforeseeable events

In closing, project execution (or implementation) is the phase in which the plan designed in the prior phases of the project life is put into action. The purpose of project execution is to deliver the project expected results (deliverable and other direct outputs).

Project execution requires the Lean and Agile PM to adhere to all the standard rules of traditional project management as well as staying flexible and keeping employees motivated. One simple option to keep things on track without stress is to make sure there is enough safety between critical events. Safety, in this case, is a term used to remind the PM to build in some small, time buffers.

Chapter 7

Monitoring, Controlling, and Closing a Project

"I knew who I was this morning, but I've changed a few times since then."

Alice, in Lewis Carroll, *Alice's Adventures in Wonderland*

As mentioned in the previous chapter, communication is key when executing the project. It is even more important when monitoring, controlling, and closing out the plan. There are a number of tools available in Lean thinking and Agile techniques (Lean and Agile) that will help the traditional project manager work through these phases. One tool that can actually be created in the planning phase is the data collection plan. Information in this plan is not only useful in monitoring and controlling a project; it can be used as a review tool in closing a project. Leadership and change management are also crucial in these final phases.

The Data Collection Plan

A useful tool in communicating the plan starts in the infant stage of the project conception and is called is a data collection plan. The purpose of a data collection plan is twofold. First, it is a communication tool. Second, it is a strategy for how the right data will be collected and used. The choice and logic of the measurement must also be articulated. For example, if sigma will be used to measure the project, a short description of sigma should be included.

In many cases, this can be accomplished by examining the data themselves to check for items that seem to be out of order. A data collection plan should document the phases including pre-data collection and post-data collection. The five steps involved should include the following:

Clearly defined goals
Reaching an understanding (authority)
Ensuring the data are reliable
Collecting the data
Following through with results

In addition to data collection plans, a project manager (PM) should have some basic understanding of change management theory. There are many modern and popular books on the topic; however, most current publications on change management still rely on a few basic concepts discovered several decades ago. Learning or reviewing how people react to change on the whole can be beneficial to a PM when communicating the plan. Traditional studies in project management do address the subject, but it is still considered more strongly in Lean and Agile.

Many team members are only concerned with how their day-to-day lives will change. Giving too much information can be overwhelming. Other team members fear a loss of position or territory.

Groups need their territories, and that it is one of the ways they define themselves as groups. The concept of group definition is important to consider when addressing change.

It is important to remember that people will want to protect their territory. This is natural and should be expected. When managers or leaders use words such as reorganization and team members realize that this means a redistribution of territory, it is unreasonable to think that some problems won't surface immediately.

Change Management

Beliefs and values evolve with a company's history. They are not easily abandoned. Change that does not address or respect these values and beliefs will most likely not be successful. In order to function in a change management role, it is important to do the following:

Stay current on the organization's mission, policies, and plans

Be prepared to communicate the mission, policies, and plans

Act as a buffer between executive management and team members during stressful mergers, layoffs, or changes in direction

Although change is an inevitable reality, framed correctly, it is possible to return to the past for inspiration. A mistake in basic change management is to discount the old way of doing things entirely. This isn't necessary and can be counterproductive.

Team members cannot be considered malleable material when it comes to change management initiatives. It is impossible to handle people like a sculptor molding clay into various forms. There is pushback and resistance even when the change is ultimately a positive one.

A surprising number of employee relations issues can be directly traced back to how well change was presented or handled within a group. There was a time when change was temporary and always followed by a longer period of stability. Now, change is continual and does not allow team members time to regroup and accept the change before even a newer change is imposed.

The way an employee approaches the thinking process can determine how well that employee will adapt to change. Thinking outside the box, thinking analytically, and thinking holistically are all indicators that the employee will be able to adjust quickly.

Individual skills and competencies can position some individuals to accept change better than others as well, and these include the following:

Technical ability

Understanding project methodology

Ability to create solutions

Capability of forming partnerships

Without natural competencies and skills, individuals exposed to change may benefit by learning about and implementing systems thinking. This thinking involves five easy steps:

Stating the problem

Telling the story

Identifying the key variables

Visualizing the problem
Creating loops

Stating the problem is the first step in almost every methodology. In areas dealing with change, it is crucial. Having the employee put the problem in story form helps the employee identify more closely with the issue. Variables are components in the story that may change over time. Variables may include things such as a change in management. Visualization of the story in graphic form sometimes helps detect the change or behavior necessary. Finally, taking the story and illustrating which factors influence other factors is called looping. There are two types of loops:

Reinforcing
Balancing

Reinforcing loops are self-fulfilling prophecies, either positive or negative. Balancing loops keep things in equilibrium.

A collaborative approach to change almost guarantees high participation, strong commitment, and the creation of a reasonable standard that may be measured for results.

A mandate from executive management is often given to the human resources (HR) department. This may include delivering training and composing communications relative to change management. There are a number of things that can go wrong with this approach, but the worst culprit is hurried communication.

A PM is expected to take an active role in the change management process. This can be challenging for a PM because the executive leadership team may very well expect a metamorphic change to occur overnight. In an effort to hurry things along, some PMs have been tempted to use e-mail when a face-to-face meeting would be more appropriate.

A popular meaning of the term managing change refers to making changes in a planned and systematic fashion. Rather than allowing change to occur naturally, and often randomly, change management assumes that it is possible to introduce planned change and steer its development.

It is important that PMs learn to embrace and demonstrate good change agent skills. Internal changes may be triggered by events originating outside the organization—environmental change—which are out of the manager's control. Implementing new methods and systems in an ongoing organization takes patience.

It is helpful to think of managing change in the same light as basic problem solving because it is a matter of moving from one state to another just as problem solving moves from the problem state to the solved state.

As a PM, it is important to understand the company's philosophy to make a positive impact on change. It is important that a PM's objectives embrace and align with the philosophy of the company. Some costly mistakes can be made by a PM who has not been involved in the change management process. Areas to watch include the following:

No systematic plan
Under-communicating the vision
Declaring victory too soon

One of the best-known philosophies was that every GE business had to be No. 1 or 2 in its market, according to Jack Welch. Otherwise, it should be fixed, sold, or closed. Introducing change that does not mesh with the philosophy of the company will not be successful.

William Bridges is the author of the two best-selling books, *Transitions* and *Managing Transitions*. Bridges makes a distinction between change and transitions and states that it isn't the changes that do managers in; it is the transitions.

Bridges believes that change is situational, such as moving to a new home. He describes transition, on the other hand, as the psychological process people go through as they internalize and come to terms with the details of the new situation. Bridges divides transition into three phases:

Ending phase
The neutral zone
The new beginning

Bridges asserts that if managers don't let go (ending phase), then it is impossible to move through to the neutral zone. Moving into the neutral zone is necessary if managers want to reach the final stage, the new beginning. These stages have also been referred to as unfreezing, changing, and refreezing.

Bridges is considered the pioneer of change management theory. His concepts were expanded on and adopted by Wharton's Center for Leadership and Change Management. They include the following:

Stimulate basic research and practical application in the area of leadership and change.

Foster an understanding of how to develop organizational leadership.
Support the leadership development agendas of the Wharton School and
University of Pennsylvania—this means consistently updated data cre-
ated by qualified groups of scholars.

In more traditional textbooks, change management has three basic areas
that should be examined: the actual task of managing change, the new body
of knowledge that must be delivered, and how the change will impact the
professional practice.

Change management requires political, analytical, people, business, and
system skills. Organizations are social systems. Without people, there can
be no organization. Guessing won't do. Change agents must learn to take
apart and put together components, considering the financial and political
impact.

There is no single change strategy. When developing a change strategy,
it is important to consider that successful change is based on the communication of information. Redefining and reinterpreting existing norms
and values and developing commitments to new ones is also essential for
success.

Too often, a PM communicates the plan by simply reporting or posting
it. It is important that a PM takes the time, even on smaller projects, to take
a lesson from Lean and Agile and place a premium on understanding the
voice of the customer, voice of the employee, voice of the process, and voice
of the business when communicating the plan.

Making Communication Easier

There are a few things that can be done to make communication go more
smoothly. Some of these items are better addressed in the planning phase,
but often the people planning the project are not on the same team as those
executing the project, which is why they are covered in this section.

1. Let the project team determine, if possible, what tools they would like
   to use.
2. Communicate early and often.
3. Leave room on your work breakdown structure (WBS) for adjustments.
4. Watch for opportunities and threats.
5. Keep a sense of humor.

Specific Activities

During this phase, the specific activities include, but are not limited to, the following:

Managing and tracking decisions
Managing and tracking action items
Execute and revise project schedule
Manage risk

Managing and Tracking Decisions

The PM is responsible for ensuring decisions that need to be made are made before they impact the project and decisions are placed in the repository of record for future reference.

In traditional project management, a decision-tracking log is often common practice. This is a good start because a decision log is basically a spreadsheet that will provide the following:

A reference for the decision
Date decision made
What was agreed and why
Who agreed to it
Where you can find more information or supporting documentation
   (optional)

In Agile decision making, the emphasis is more on making the decision than the document. However, understanding how Agile approaches a problem may make the decision log more robust.

With that in mind, here is the basic structure of an Agile framework that should help you better handle tough business decisions:

Start with eliminating irrelevant information or choices that don't align with your project.
Anticipate the presence of decision anxiety by committing to a time frame for taking the first step. Limit how much time you can research or contemplate a choice before committing.
Set a date on the calendar to review key decisions weekly.

Be deliberate about when you involve outside feedback. Bring in collaborators and partners.

Practice being present with the consequences that come from your decisions, both good and bad.

Managing and Tracking Action Items

The PM is responsible for ensuring that tasks too small to appear in the project schedule are recorded and completed.

In traditional project management, a list is kept or items are added to the WBS. This can be enhanced by using a Gantt chart to visually see when tasks are being finished. Obviously, keeping your team on track is a lot more difficult than recording the results.

Daily Scrum meetings, which are used in Agile, are meetings in which participants are asked on a daily basis to report the following along with their teams:

1. What did you do yesterday?
2. What will you do today?
3. Are there any impediments or obstacles in your way?

By focusing on what each person accomplished yesterday and will accomplish today, the team gains an excellent understanding of what work has been done and what work remains. The daily Scrum meeting is not a status update meeting in which a boss is collecting information about who is behind schedule. Rather, it is a meeting in which team members make commitments to each other.

These are a few examples of impediments or obstacles:

1. I can't get the X group to give me any time, and I need to meet with them.
2. I can't get the vendor's tech support group to call me back.
3. I need help debugging a problem with X.
4. I still haven't got the software I ordered a month ago.
5. I'm struggling to learn X and would like to pair with someone on it.
6. My X broke, and I need a new one today.
7. Our new contractor can't start because no one is here to sign her contract.
8. The department VP has asked me to work on something else "for a day or two."

Execute and Revise Project Schedule

Keep the project schedule updated by obtaining status on project tasks and updating those tasks in the project schedule. The project schedule should be monitored and updated regularly.

This is the same for anyone working in project management. It is about frequent and valid communication. The advantage to Lean and Agile communication is that there is an underlying theme that input from the team is important and people are encouraged to ask questions.

Manage Risk

An initial list of risks and management approaches are identified in the project charter. The project manager must monitor the risk list, identify any that have become issues, and implement the contingency plan identified in the project charter.

Typically, in traditional project management, a risk register and/or issue log is maintained. Just like a decision or issues log, this is helpful in not only managing the project but identifying best practices as well.

Lean Six Sigma offers a number of tools to manage risk. It also expands on the concept of what risk is. Risk isn't always concentrating on what things could go wrong. Risk examines the internal environment, such as the following:

Business vision
Leadership
Ethical issues
Organizational commitment to human capital

When exploring risk, setting objectives and risk response need to be discussed with the team. Additionally, the methods used to control and monitor activities need to be fully understood. Some tools include the following:

Cause-and-effect matrix
Control charts to track the process
Failure mode effects analysis (FMEA)

Risk identification, monitoring, and resolution are key tools to successful completion of a project. Part of controlling a project during the execution phase is to have an established risk management process. This process is begun as part of project planning (see risk management subsection of the project planning phase section) and is kept current until project closeout. The key elements to this process are the following:

Creating a central repository for risk information and associated documentation of risk items and resolution strategies
Summarizing information on a risk form
Including a risk summary in the regular status meetings
Providing a consistent and ongoing evaluation of risk items and development of risk strategies
Identify the risk
Evaluate the risk
Define a resolution

In all cases, risk management is an iterative process that is performed throughout the project.

The initial list of risks that begins with the project will evolve over time. To ensure that new risks are added and resolved risks are eliminated, risk identification meetings should be held.

Cause-and-Effect Matrix

An easy but powerful cause-and-effect matrix is the fishbone diagram. This tool is useful when trying to determine root cause or uncover a bottleneck. It is also helpful when trying to reason out why a process isn't working.

Step 1: First, write down the exact problem you face. Where appropriate, identify the problem, who is involved, and when and where it occurs. Then, write the problem in a box on the left-hand side of a large sheet of paper and draw a line across the paper horizontally from the box. This arrangement, looking like the head and spine of a fish, gives you space to develop ideas.

Step 2: Work out the major factors involved. Next, identify the factors that may be part of the problem. These may be systems, equipment, materials, external forces, people involved with the problem, and so on.

Step 3: Identify possible causes. Now, for each of the factors you considered in Step 2, brainstorm possible causes of the problem that may be related to the factor.

Step 4: Analyze your diagram. By this stage, you should have a diagram showing all of the possible causes of the problem that you can think of at that particular time (see Figure 7.1).

Control Charts

A control chart is a graph used to study how a process changes over time. Data are plotted in time order. A control chart always has a central line for the average, an upper line for the upper control limit, and a lower line for the lower control limit. These lines are determined from historical data (see Figure 7.2).

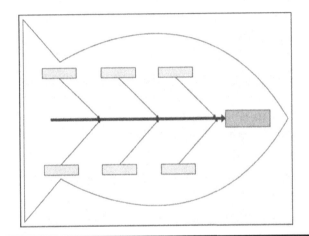

Figure 7.1   Fishbone template.

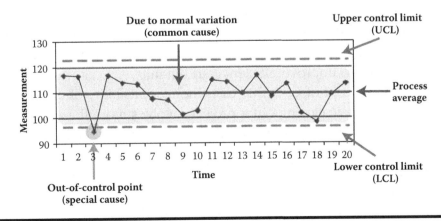

Figure 7.2   Failure mode effects analysis (FMEA).

One of the steps in FMEA is establishing a risk priority number (RPN). This is a numeric assessment of risk assigned to a process or steps in a process as part of FMEA in which a team assigns each failure mode a numeric value that quantifies the likelihood of occurrence, likelihood of detection, and severity of impact. It is a number that typically looks at tables associated with severity, occurrence, and detection as a composite of the true risk. Values in each category range typically from 1 to 10 but can range as high as 1 to 1,000.

Tools

A popular tool to monitor the WBS is the Gantt chart. This chart gives a graphic snapshot of where your project stands as far as the completion of certain activities and tasks.

Lean thinking and Agile techniques (Lean and Agile) take a more serious approach to closing projects than traditional project management (traditional). This is due the concept of continuous improvement. Lean and Agile take a special interest in documenting lessons learned and how the overall plan may have changed or been modified.

To be fair, project management has always promoted that the definition of a project is something with a beginning and an end. Lean and Agile generally believe in the idea of continuous improvement (CI). CI focuses on either how to make the current project better, faster, and more cost-effective the next time a similar project is done, or how to make the existing project more robust.

CI models include models such as plan-do-check-act (PDCA), but other models, such as Define, Measure, Analyze, Improve, and Control (DMAIC), are also in the mix. Kaizen is also known as CI and is a long-term approach to work that systematically seeks to achieve small, incremental changes in processes in order to improve efficiency and quality.

Because the concept of CI is that CI is the responsibility of every worker, not just a selected few, closing out a project involves more than the customary activities done when closing out a project such as the following:

Formal sign-off
Releasing resources
Updating paperwork
Handling any procurement issues

These activities are important and well-respected by Lean and Agile; however, many things that are considered nice to do are not required in traditional project management. Lean and Agile consider these activities essential:

Celebrate the success
Talk about lessons learned
Record best practices
Create standard work

Celebrate the Success

Working on projects is often hard work requiring many hours of dedication. This is why it is important to recognize team contributions to making any project a success. Honoring success can happen at every level of the operation. The annual performance bonus, the uplifting speech at the holiday party, the media release broadcasting news of a major corporate achievement—these are all fine ways to recognize success. But it's also important to entrench recognition of a job well done on the small scale, too. In general, celebrate project success because it is motivating and produces happy chemicals for the employees. Dopamine is released into our brain when we anticipate achieving something or we do achieve it, and it feels good, so we want more. A good point to remember is you have to recognize your own success if you want other people to as well.

Talk about Lessons Learned

Talking about lessons learned is a good team-building activity as well as a useful project management tool. Organizations gain information through experience. This experience can be relevant to future projects or organizations and should be retained for future reference.

Identification, tracking, and consistent communication of lessons learned can save organizations time and money when embarking on similar efforts by preventing recurrence in other areas of the business or in other projects.

The importance of lessons learned is how the lessons may be used as tools to improve project management efforts moving forward.

The items covered during this conversation should include, but not be limited to, the following:

Identifying the event and subsequent lesson learned

Documenting the lesson learned and associated events

Acting to resolve or mitigate the situation

Communicating key information to relevant stakeholders: when the event occurred, as it is mitigated, and post-project close out

Maintenance or tracking of the lesson learned for future reference during other projects or by other personnel

Record Best Practices

A best practice is a technique or methodology that, through experience and research, has proven to reliably lead to a desired result. It is similar but different than standard work, which is discussed next. Usually, a best practice can be applied or modified to work with more than one project or activity. Software is available to help collect and articulate best practices. Most Lean and Agile practitioners just use a spreadsheet or simple database.

Create Standard Work

Developing standard work is one of the more difficult Lean Six Sigma disciplines; however, if efficiently developed, it should allow virtually anyone to perform the work without any variance in the desired output. It makes future work more efficient and includes a detailed definition of the most efficient method to produce a product (or perform a service) at a balanced flow to achieve a desired output rate. It is something that can often be created at the end of a project.

Why Implement Standard Work?

Implement standard work to make it possible to identify and eliminate variations in operators' work and to sustain gains achieved from improvement activities. It may also be used to provide a baseline for future improvement.

Elements of Standard Work

Total labor time
Takt time
Cycle time work
Sequence
Job steps
Machine time
Standard work in progress (WIP)
Desired output

How to Develop Standard Work

Engage employees in the development process
Encourage team members to collaborate
Be realistic
Guide through the creation and provide final approval
Design with the intent to make problems visible
Capture takt time of the process

Project management consists of five process groups, namely initiating, planning, execution, monitoring and control, and closing. The project closure phase consists of the processes that are performed to officially finish and close all the assignments in a project or phase. The concluding action in this group will usually include the approval and transfer of the project deliverables to the user or customer. It is important that all actions that are included in the closing group are performed thoroughly because only then will a project be considered as closed. The essence of this group is that all the stakeholders agree that the project has met its assigned objectives, and additional charges will not be assigned to the project. Furthermore, the staff and other resources can be assigned to other projects or relieved if not required. Lean and Agile expand on this by focusing on lessons learned and establishing best practices.

Chapter 8

Applying Lean and Agile Techniques to Project Management Areas of Knowledge Promoted in the PMBOK®

Most project managers (PMs) believe there are key knowledge areas that PMs should be aware of and be able to practice. As a reminder, the knowledge areas as identified in the PMBOK® guide are the following:

Project integration management
Project scope management
Project time management
Project cost management
Project quality management
Project human resource management
Project communications management
Project risk management

Almost anyone working in project management will agree that these are crucial areas. However, in addition to the ones captured by the PMBOK® guide, other important areas that could be thought of as knowledge areas include the following:

Buy-in
Team management
Leadership

This work takes the approach that the above three bullet points are embodied in the eight knowledge areas as presented by the PMBOK® guide and are addressed in whole or in part either in this chapter or other areas of this book.

There are a few tools that the PM may already utilize mentioned in this section. However, even tools known to the PM seem to have a more robust use and description when seen through the Lean and Agile lens. The following also considers that sometimes it isn't about the tool that Lean or Agile can provide, but rather the direction and mindset.

Project Integration Management

Integration management is a collection of processes required to ensure that the various elements of the projects are properly coordinated. It involves making trade-offs among competing objectives and alternatives to meet or exceed stakeholder needs and expectations.

For integration management to be effective, you need to get buy-in from key stakeholders and team members. Getting buy-in from the get-go will ensure that your project receives the support and funding needed for it to be successful.

To get buy-in, start by creating a project charter and a preliminary scope statement. Lean and Agile tools that may be useful in assisting in this area include the following:

Continuous integration (CI)
Value stream mapping
Takt time
Fishbone
Poka-yoke

Continuous Integration

In Agile programing, through the use of various products that promote Agile software, isolated changes are immediately tested and reported on when they are added to a larger code base. The goal of CI is to provide rapid feedback so that if a defect is introduced into the code base, it can be identified and corrected as soon as possible. Thinking of this approach as a concept applied to service models is valuable if the PM is anticipating many decision points in the project.

Value Stream Mapping

The focus of this tool is to identify and eliminate any non-value activities and to reduce the wait time between consecutive steps wherever possible. Non-value activities are those activities or tasks that do not directly contribute to the process. In other words, looking at the time the customer placed the order to when the order was delivered what things were absolutely necessary? Whereas a process map shows every step, a value map only shows the things that must happen for the product or service to be delivered to the customer.

Takt Time

Takt is a German word that can be roughly translated as "beat." Takt time is the rate at which a completed project needs to be finished in order to meet customer demand. For processes involving cycle times, such as manufacturing or incident management, the as-is cycle time can be captured.

Fishbone and Five Whys

These two concepts are discussed in-depth in Section III of this book, but as a quick explanation, both are root cause analysis tools. The fishbone diagram takes the approach of putting the problem in the head of a fish diagram and using the fish bones to depict major contributors. The Five Whys is a verbal exercise in which the question "why" is asked five times.

Poka-Yoke

A Japanese phrase meaning mistake-proofing, poka-yoke can be used to tune process steps and also when designing a new system. Here are step-by-step instructions:

1. Identify the operation or process based on a Pareto.
2. Analyze the five whys and understand the ways a process can fail.

3. Decide the right poka-yoke approach, such as using a shut-out type (preventing an error being made) or an attention type (highlighting that an error has been made). Poka-yoke takes a more comprehensive approach. Instead of merely thinking of poka-yokes as limit switches or automatic shutoffs, a poka-yoke can be electrical, mechanical, procedural, visual, human, or any other form that prevents incorrect execution of a process step.
4. Determine whether a contact—use of shape, size, or other physical attributes for detection, such as constant number in which an error is triggered if a certain number of actions are not made or a sequence method in which a checklist is used to ensure the completion of all process steps—is appropriate.
5. Trial the method and see if it works.
6. Train the operator, review performance, and measure success.

Project Scope Management

Scope refers to the detailed set of deliverables or features of a project. These deliverables are derived from a project's requirements. Project scope is the part of project planning that involves determining and documenting a list of specific project goals, deliverables, tasks, costs, and deadlines.

This is an area that differs between Lean and Agile as much as it does between Agile and traditional project management. In Lean and traditional project management, the scope has been defined in a specification document, such as a project charter.

Whereas in an Agile project, as with traditional approaches, the scope defines what is to be designed and delivered, there are several differences. For example, Agile is concerned with things such as timing, attitude, and high-level requirements. Agile scope is a continual conversation. Agile change management is used to facilitate change. Agile change management involves changes to the release plan and changes to the time box plan discussed in Chapter 3: Agile Comprehensive.

Scoping is more of an art than a science as it requires understanding the customer's needs as well as the technical ability of the company. Some tools in Lean that are discussed more fully in Section III of this book include the following:

Critical to quality (CTQ)
Voice of the customer
Supplier, input, output, process, customer (SIPOC)

Project Time Management

Time management in general falls to the PM; however, project time management is considered differently than the typical meaning. Project time management involves identifying and scheduling different components of the project management sequence that are necessary for project deliverables to be accomplished on time. Things such as estimating activity duration and watching for barriers that may delay the project are included. This is one area in which both Lean and Agile can benefit from adopting traditional time management approaches and practices.

Project time management is dynamic and often requires input from the team. Therefore, some of the communication tools used in Lean may be beneficial. These include items such as a data collection plan outlining who will collect the data and when the data will be collected. Another Lean tool mentioned earlier and discussed in Section III is the SIPOC model. This model helps identify all the human and nonhuman resources involved in the project to make sure that timelines are managed around those people and things involved.

Schedule control is an important factor. Agile uses the Scrum meetings to manage the day-to-day activities so there are no surprises.

Project Cost Management

Project cost management (PCM) is a method that uses technology to measure cost and productivity through the full life cycle of enterprise-level projects. PCM encompasses several specific functions of project management, including estimating, job controls, field data collection, scheduling, accounting, and design.

This is probably the best news for the traditional project manager. No matter how innovative project management is, the way we calculate and manage financial information is the same.

Lean tools focus on reducing waste, which, by default, will reduce cost, giving the PM more flexibility. A company with effective cost-reduction activities in place will be better positioned to adapt to shifting economic conditions.

The most helpful thing from Lean in this area is the TIM WOODS model. This is also known as the eight areas of waste. Concentrating on each one of these areas, if they apply to the specific project, will allow the PM to determine if waste in the project is present and can be reduced or eliminated. TIM WOODS is a way to remember the eight areas of waste.

T: Transport, moving people, products, and information
I: Inventory, storing parts, pieces, and documentation ahead of requirements
M: Motion, bending, turning, reaching, lifting
W: Waiting, for parts, information, instructions, or equipment
O: Overproduction, making more than is immediately required
O: Over processing, tighter tolerances or higher-grade materials than are necessary
D: Defects, rework, scrap, incorrect documentation
S: Skills, underutilizing capabilities, delegating tasks with inadequate training

Remember, Agile welcomes change, and that applies to budgeting as well. The most helpful Agile practice in relationship to cost management is that it combines cost and schedule performance on one sheet and helps control the entire project.

There are also Agile ways of processing earned value management (EVM). These are discussed in Chapter 3: Agile Comprehensive.

Project Quality Management

In Lean, quality management is the act of overseeing all activities and tasks needed to maintain a desired level of excellence. This includes creating and implementing quality planning and assurance as well as quality control and quality improvement. It is also referred to as total quality management (TQM).

Project quality management is all of the processes and activities needed to determine and achieve project quality, so the definitions have a lot of synergy.

The main difference between quality in its general meaning and that of project management quality is that Lean's definition of quality encompasses the entire organization, and quality initiatives and practices are on an enterprise-wide level.

Lean, Agile, and traditional project management all agree that the true test of quality rests on three major concepts: customer satisfaction, prevention, and CI.

In many ways, traditional project management has the customer satisfaction concept area covered better than Lean or Agile. However, Lean excels in prevention, and Agile excels in detection—being able to change before an error is made. Both Lean and Agile are committed to continuous improvement as it is part of their methodologies whereas traditional project management does not typically sponsor improvement or even sustainability models.

Agile favors making small incremental changes and leaders who walk the talk. Lean favors fire prevention as opposed to fire fighting. In Dr. W. Edwards Deming's 14 points, he called for the "constancy of purpose for continual improvement of products and service to society" (see Chapter 3).

Both Lean and Agile promote shifting to a long-term mindset as opposed to focusing on a single project. Managers are often focused on whether they're going to meet their monthly or quarterly targets, and it can be very difficult to prioritize improvements that will only make an impact over the longer term. As a result, CI is as much about mindset as it is about actions.

Project Human Resource Management

Again, to the general community, the customary term human resource management means something a little different than in project management. When planning human resource management, the first thing is to identify all the project roles and responsibilities. Documenting the reporting relationships and the staffing management plan are key in the planning process. Project roles are roles taken up by individuals or groups within or outside of the project itself. Project human resource management is focused on the project team and its organization, management, and leadership. It is defined as the processes that organize, manage, and lead the project team.

Again, due to Scrum meetings, which are daily meetings, Agile techniques are very helpful in knowing the exact parameters of the team and the project.

Lean leadership includes an entire workshop, taught by many vendors, that concentrates on leadership and communication skills along with using communication tools. Many of the tools highlighted in the next part of this book in Chapter 12: Making the DMAIC Model *Leaner* and More Agile: Analyze can also be applied. Major concepts include, but are not limited to, the following:

1. Acknowledge that communication is a critical part of your job. You must communicate. Even if you're quiet, your silence and your actions will send messages.

2. Recognize that information overload is a significant barrier to effective communications. So be mindful about what you say and do to ensure you're sending compelling messages. You want to cut through the clutter and support your strategic intent and actions. Spare the air on non–mission critical issues.

3. Try to meet face to face regularly. It allows people to hear you talk. It should also give them an opportunity to ask questions, seek clarification, and share opinions with you. In other words, to ensure your face-to-face communication is effective, make it two-way.

4. Be accountable by measuring, adjusting, and reassessing.

5. Encourage people to speak in an organizational context, making it safe and comfortable for individuals, especially those in less powerful positions, to speak up. This ranges from raising issues to questioning authority.

Project Communications Management

A communications plan, in project management, is a policy-driven approach to providing stakeholders with information about a project. The plan formally defines who should be given specific information, when that information should be delivered, and what communication channels will be used to deliver the information.

Unlike traditional Waterfall methods, roles and responsibilities in Agile teams are distributed equally among all the members on a project. This makes for a flatter environment. This makes communication plans drastically different that those in traditional project management, but the positive thing is that at any given point in time, if the Agile communication is being run true to the theory, a Scrum master/project manager knows exactly where the project is as far as time and cost baselines.

Lean, similar to traditional project management, favors customary project communication plans and depends primarily on the project plan done using a work breakdown structure (WBS). Whereas the WBS is supplemented by short or summarized communication, the basis of the communication is the documented project plan.

The communication plan is still different than the WBS or what some environments may call the rollout plan in the sense that it communicates the roles of the individuals and the plan's purpose. This normally would include goals and objectives as well as the kick-off plans.

The communication roles are the following:

Project sponsor
Project manager
Leadership/management team
Steering team
Project lead
Project team member

Ways to communicate the message include the following:

Meeting summaries
Status reports
Newsletters
Formal presentations
Surveys
Internet/intranet web pages
Informal small group meetings
Brown bag lunch workshops

Project Risk Management

Project risk management is about identifying and managing events or conditions that could have a negative impact on the project objectives. There are five areas to consider:

1. Identification
2. Analysis
3. Evaluation
4. Treatment
5. Monitoring and reviewing

In traditional project management, attention is paid to documenting these areas and watching via a manual or computerized checklist. Risk management issues are recorded in a log and monitored.

In Lean, more attention is given to poka-yoke or error-proofing. Basically, the system concentrates on building risk management into every step of the process. Defects occur when the mistakes are allowed to reach the customer, so this system eliminates issues on the spot. In other words, something that goes wrong isn't handled the next day but in real time whenever possible.

Another train of thought used by Lean and Agile is the theory of constraints (TOC). This theory claims that, because there is always at least one constraint (obstacle) in a project, attention should be dedicated to brainstorming the events and then taking each constraint and exploiting it so that it becomes an asset instead of a liability.

A constraint is anything that prevents the system from achieving its goal. There are many ways that constraints can show up, but a core principle within TOC is that there are not tens or hundreds of constraints. There is at least one, but at most, there are only a few in any given system. Constraints can be internal or external to the system. An internal constraint is in evidence when the market demands more from the system than it can deliver. If this is the case, then the focus of the organization should be on discovering that constraint and following the five focusing steps to open it up (and potentially remove it). An external constraint exists when the system can produce more than the market will bear. If this is the case, then the organization should focus on mechanisms to create more demand for its products or services.

There are often internal constraints, for example, equipment limitation, untrained employees, or prohibited policies.

In closing, the Project Management Body of Knowledge – PMBOK® supports a number of knowledge areas. These areas include integration, scope, time, cost, quality, human resources, communication, and risks. There are tools in the Lean Body of Knowledge that help identify issues and implement templates and solutions in these areas. The Agile way of managing projects allows for team members to openly discuss these important topics. Both Lean and Agile support the proactive concept of identifying potential issues or constraints that could hinder the project.

Chapter 9

A Leaner, More Agile Approach to the Project Management Life Cycle: SSD Project Life Cycle™

"And how many hours a day did you do lessons?" said Alice, in a hurry to change the subject.

"Ten hours the first day," said the Mock Turtle: "nine the next, and so on."

"What a curious plan!" exclaimed Alice.

"That's the reason they're called lessons," the Gryphon remarked: "because they lessen from day to day." —The Mock Turtle's Story

Alice, in Lewis Carroll, *Alice's Adventures in Wonderland*

A leaner, more agile approach to most accepted life cycles dealing with project management is the one created by Simple, Smart Decision-Making (SSD) and is called the SSD Project Life Cycle™. This easier way of thinking is especially useful for small projects. Some project management models over-complicate the steps necessary and cause the project to slow down.

The premise for the SSD Project Life Cycle™ (SSD-PLC™) is to break projects down into four basic areas:

1. Selection
2. Planning
3. Execution
4. Close out

Here is a quick understanding of the simplified framework.

Selection

The following things would occur in the Selection phase of the SSD-PLC™.

- Define the purpose
- Determine the strategic fit
- Define the objectives
- Draft a scope statement
- Discuss time and budget concerns

Sometimes the most crucial criteria and information are lost prior to the planning. Planning would be easier if the ground rules were established first. Phase One of the SSD-PLC™, Selection, would establish these parameters. In some cases, the Selection Phase might determine that a better or more cost-effective project should be selected. The advantage of doing or not doing a particular project is clearly determined before much effort is put forth. The end result of this phase would be a clear project charter document.

Planning

The following things would occur in the Planning Phase, or Phase Two, of the SSD-PLC™. This phase would begin with the business case from the Selection Phase. This phase would begin with the agreement, in place, that we would be moving forward with the project.

- Finalize scope
- Outline the deliverables

- Create a risk schedule
- Capture the quality requirements and document
- Plan for the constraints
- Draft the plan based on scope, deliverables, risk, quality requirements, and constraints
- Develop cost and schedule timelines
- Select the team members and assign resources to task
- Finalize the project plan

When a plan is drafted and reviewed by various employees including the sponsor, the potential team, and, in some cases, the client, it is easier to develop baselines (time and cost) and to select appropriate team members. In traditional project management, many times, the team is selected first without an understanding of the time and cost projections.

The end result of this phase would be a solid, agreed-on, project plan. The suggested format is the Work Breakdown Structure (WBS) supported throughout this book.

As a reminder the WBS breaks the team's work down into manageable sections. It is an outline that first captures the key tasks and labels these tasks: 1, 2, 3, and so forth. Then under each task additional items are captured as 1.1, 1.2, 1.3. A subtask of Item 1.1 would be 1.1.1 and next 1.1.2.

Execution

The following things would occur in the Execution Phase, or Phase Three, of the SSD-PLC™. This phase would begin with a ratified project plan in the WBS.

- Communication of the plan to all individuals associated with or impacted by the plan
- Review of the timeline with key resources
- Monitor the financials
- Adhere to the quality plan
- Reporting/meetings as needed until the project is fully executed

The end result of this phase would be fully executed and draft documents of the project.

Close Out

The following things would occur in the final phase, Phase Four, Close Out.

- Finalize any documentation from the drafts produced in Phase Three, Execution
- Post-Implementation Review
- Feedback and recommendations—management, team, and customer (internal or external)
- Record best practices
- Notify anyone who participated or was impacted by the project of the Close Out
- Celebrate

The end result of this phase would be a published report documenting the bullet points.

The SSD Project Life Cycle™ has a number of benefits in addition to simplifying the process. It ensures the risks are low and the message is appropriate and understood at all levels of project engagement. This provides direction and safety for the team and an uncomplicated road map. The most common reasons that projects fail are people-related. The SSD-PLC™ supports communication.

Chapter 10

Making the DMAIC Model *Leaner* and More Agile: Define

DEFINE, Measure, Analyze, Improve, Control

When Simple, Smart Decision-Making Inc. (SSD) first designed the *Leaner*™ the idea was to provide an easier framework for Lean Six Sigma students. The results however applied directly to project managers wishing to incorporate some Lean practices into their regular projects. This chapter covers the first step in the Define, Measure, Analyze, Improve, Control (DMAIC) process but using easier terminology. The value of learning about Define is that at the beginning of all projects these tools can be utilized regardless if the intention is to make the project a DMAIC effort.

The major objective in the Define phase is to identify the process improvement goal and document the supporting information. What problem needs to be solved? What process improvement can be made? Is it reasonable to take this specific existing process and invest the time in making that process better, faster, or more cost-effective? If yes to any of those questions, then how can I best communicate the problem and show a need for improvement?

The DMAIC model is responsible for determining the next steps that will lead the project manager to how the problem will be solved. The Define phase must define the problem and establish confirmation that the process improvement will have value to the project sponsor.

The method normally used to determine if the time and effort are worthwhile is called return on investment (ROI). Because one of the primary functions of the DMAIC model is to provide reasonable solutions, in the Define phase, the project manager is working with projected ROI. Projected ROI is determined by estimating what overall benefit might be realized if the existing process was improved.

This first phase is also about securing an agreement on how well the current process is working. This is normally proven by designing a current process map. Crucial insight is often gained by simply mapping the current process although process improvement solutions are not recorded until the end of the Analyze phase of the DMAIC model.

Key tools for the Define phase include the following:

Process map
Project charter
Strengths, weaknesses, opportunities, threats (SWOT) CTQ definitions
Stakeholder's analysis
Supplier, input, process, output, customer (SIPOC) diagram
Quality function deployment/design (QFD)
DMAIC WBS—project tracking tool
Affinity diagram
Kano model

The Define, Measure, and Analyze phases of the DMAIC model are somewhat creative, making use of tools that inspire the team to be inventive in finding creative solutions. As long as the general objective is met, the PM has a great deal of flexibility.

In the Define phase, for example, there are only two rules to qualify as a DMAIC project. The PM must have, first, a map of the current process and, second, a project charter before leaving the Define phase. With these two tools, everyone should understand what they are working on. Certainly, all the other activities suggested in the Define phase will lead to greater project success; however, these two documents are the only hard and fast requirements in the Define phase.

Process Mapping

Process mapping can be simple or complex. The only rule is that it must reflect the current state: what is the current daily process. Each detail, even unusual or unexpected things, needs to be recorded. When a process map is provided, the PM should still walk the process to verify that the map reflects the actual current process since the last map was produced. It is surprising how many PMs try to improve a process without fully understanding the current state.

A process map is often displayed with visuals using flowcharting symbols. The map may also be visualized by using a bulleted list or a list that reflects Step 1 in the process, Step 2 in the process, and so forth.

Process mapping is simply taking the steps in the process and applying graphic symbols. Often flowcharts are designed on the basis of what is supposed to be in place rather than reality. Post the process flowcharts in an area large enough to allow clear visual representation where everyone involved can review the information. Special considerations will need to be made for information involving confidentiality or security issues.

One type of process map is a value stream map (VSM), also discussed in the Analyze phase. A VSM is a paper-and-pencil tool that helps the viewer understand the flow of material and information as a product or service makes its way through the value stream. If there is a process already in place, an effort to identify any "hidden" processes should be made. The VSM combines several types of charts, such as swim lane and PERT, to visually show the process flow (see Figures 10.1 and 10.2).

Project Charter

The main purpose of a project charter is documentation for the PM, sponsor, and team. A project charter is a document that records the purpose of the project along with additional information including items such as why the project is being initiated and who will be working on the project. In a typical project management scenario, often a project plan is prepared. This would include a list of the tasks to complete and an estimated idea of the time, costs, and resources necessary to complete the project.

A process improvement project charter is more like a proposal. It lists the process to be improved, why it should be improved, and the projected benefit. The objective is to be given permission to work on the project and to ensure that all parties involved understand the specific process improvement that has been targeted.

Sometimes a basic project charter has already been published before entering the DMAIC process. If this is not the case, the project charter must be fully developed in the Define phase. As mentioned earlier, it is not

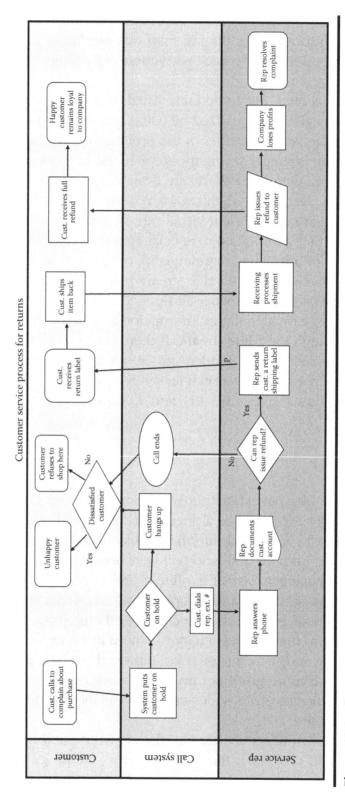

Figure 10.1 Process map.

Define: Current process map for workout routine

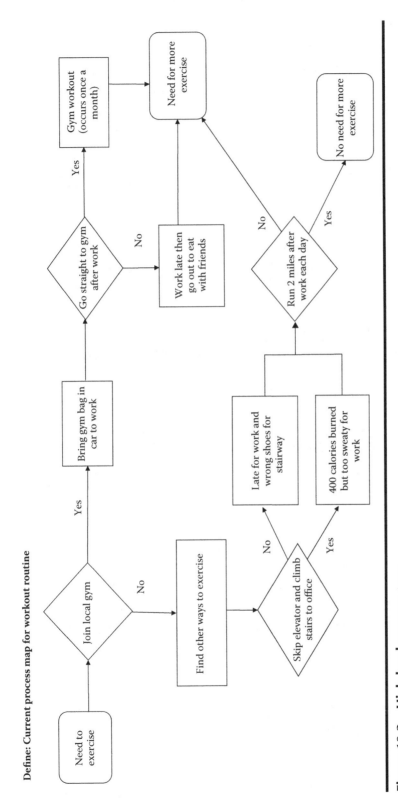

Figure 10.2 High-level process map.

possible to leave the Define phase and enter the Measure phase without a clearly defined project charter.

Most project charters begin with recording the problem statement. The problem statement may be a real problem that needs to be solved, or the problem statement may be an activity that needs to be accomplished. A vague project definition leads to unsuccessful proposals and unmanageable documents. Naming a topic is not the same as defining a problem, but it is a good place to start.

A problem statement in a project charter is similar to writing a thesis statement. A problem statement is specific. Sometimes the problem statement is formulated by a steering committee, but more often it is the responsibility of the PM to develop the problem statement.

When brainstorming possible problem statements, here is a methodology that may be useful:

Make a list of everyone involved.
Find out what users consider to be the problem.
Group the problems into categories.
Condense the main categories into a problem statement.

A solid project charter aids in overall mistake-proofing efforts. Mistake-proofing is a constant theme throughout the DMAIC model. Poka-yoke is the Japanese phrase for do it right the first time and is often referred to as mistake-proofing methodology.

Before a project is selected, the PM needs to determine the criteria. In some companies, there will be a clear methodology explaining the project selection process. The selection may be based on a number of factors, such as ROI, process capability, or even green (environmental) factors. When a selection methodology is not clearly defined, a tool called SWOT analysis may be helpful.

There are a number of reasons a project charter is necessary to facilitate the DMAIC model. A solid charter will do the following:

Provide a clear statement of work.
Outline critical success factors.
Define expected benefits.
List key stakeholders.
Clarify what is expected of the team.
Keep the team focused.

Keep the project and team aligned with organizational priorities.
Name constraints and assumptions.

Project charter templates are easily available; however, if the company has an existing template it is best to use that as a foundation. Information not covered in the company's template should be added as an appendix.

SWOT Analysis

A SWOT analysis looks at quadrants to determine, via brainstorming, the strengths, weaknesses, opportunities, and threats of a project. Strengths and weaknesses may be thought of as pros and cons. Another diagram that displays the pros and cons is a force field analysis. The SWOT diagram takes on additional factors, such as threats (risks) to the project as well as opportunities or possibilities. The SWOT analysis is helpful in overall decision making (see Figure 4.3).

One of the advantages of SWOT is that the PM can determine immediately if there is a solid reason to move forward with the process improvement or if a different process improvement should be considered. It forces the PM and/or the team to clearly state what needs to be improved. Examples of existing processes that a SWOT might consider would be the following:

Improve the student enrollment system
Improve the car rental process
Improve the method to design online applications

In each one of the above examples, what would be the pros and cons about making the improvement? What would be the possible opportunities? What would be the risks?

Projects do not generally begin without a sponsor. The sponsor is the person funding the project. It is recommended that there is also a person to function as a champion.

The champion is the person who will help with issues such as change management and publicity about the project. When the problem statement is fully developed and the project funding has been secured, this would be the time to lobby for help from a senior manager who has something to gain from the project success. The SWOT analysis can be a useful tool when trying to recruit champions as it gives a quick overall picture of the issue.

Critical to Quality

In the Define phase of the DMAIC model, once the problem statement has been determined, the next step is to consider CTQs. An easy way to think of CTQs is as anything that is important to the success of the project. This makes customer requirements and expectations, by default, CTQs. However, CTQs should not be limited to only the customer and should include anything that needs to be considered for the successful completion of the process improvement. CTQ is often used as an umbrella term and could include critical to speed, critical to cost-effectiveness, or critical to success of the project.

CTQs are the key measurable characteristics of a product or process. A CTQ is usually interpreted from a qualitative customer statement. It must be an actionable and measurable business specification. CTQs are what the customer expects of a product. Discussing the process boundaries and the customer's goals is essential to success.

CTQs can be developed from a stakeholder's analysis. Stakeholders are people who will be affected by the project. A stakeholder analysis is a matrix (chart) that describes the project impact on each stakeholder. Once stakeholders are recorded, the project manager can go directly to the stakeholder and determine what is important to that stakeholder. This information will become CTQs along with the customer requirements.

The voice of the customer (VOC) is the term used to describe the stated and unstated needs or requirements of the customer. The VOC is also considered a process used to capture the requirements or feedback from the customer. Seeing things from the customer's perspective may be accomplished by direct discussion, interviews, surveys, focus groups, and even complaint logs. Other methods include warranty data, field reports, and customer specifications. The VOC is critical to the project and to determining the validity of the CTQs (see Figure 10.3).

As previously mentioned, one of the major outcomes of the Define phase is to provide clarity as to the process improvement. CTQs are very useful in this determination. CTQ knowledge can provide valuable information on how the improvement may positively impact business or department initiatives.

Define: CTQ using VOC

VOC for local tax company and marketing for the local tax company to bring in new clients.
Voice of the customer:
Accurate tax return
A brand and tax associate they trust
Great customer service
Communication before, during, and after
Can this service be completed online to save money and time?
Want value with saving or a discount for the fee
Clear idea of cost of return
Options for tax payments
Convenient time of appointment and location
Knowledge and explanation of deductions
Where to go with IRS problems
Voice of the community:
General distrust, and dislike of solicitation from marketing
Anger, distrust and resentment paying taxes to government
Lack of understanding for tax laws, including health care
Need for diversity, including ITIN and bilingual associates
Tendency to postpone paying taxes as long as possible
Voice of current local marketing process:
High cost of lost opportunity
High cost of the current process for labor and wasted materials
Office associates are not trained for marketing
There is no strategic marketing plan including securing local events
No consistent message or standard of marketing interactions
How to accurately measure the results weekly to identify success
Best practices not being shared
Need for very quick changes in marketing direction and clients, needs
Voice of the tax associates:
My primary job is completing returns, not marketing
I have no idea how or where to go for marketing
I need to make my commission bonus
The office needs me inside
Don't want to work weekends or extra evenings
Challenged within office learning new software
Giving out a discounts cuts my commission/revenue

Figure 10.3   CTQ/VOC.

SIPOC Diagram

SIPOC is a diagram that helps determine CTQs as well as ensure that all factors are being considered in the problem statement. SIPOC stands for the following:

Supply
Input
Process
Output
Customer

A SIPOC can be formal or informal but is a crucial risk management tool. It is generally used to help identify all of the people and things that should be considered before starting on a project. In some ways, it replaces the quality circles promoted by total quality management (TQM).

TQM suggested that everyone who touched the process should be involved. This level of detail proved to be very costly. By considering the areas in the SIPOC diagram, all the stakeholders that need to be considered and all resources that will be required can be determined. A SIPOC may be used as a problem-solving tool (identifying all the stakeholders in a project) or as a tool in the Measure or Analyze phase of the DMAIC (determining key data measurements). SIPOC can also assist in process mapping (see Figure 10.4).

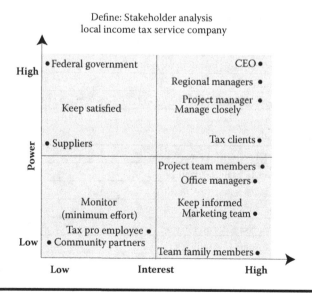

Figure 10.4   Stakeholders.

Quality Function Deployment

QFD is more complicated. The QFD portion first gathers the customer requirements. Second, it maps these requirements with the technical capabilities that will be needed. The output of QFD is often referred to as a House of Quality. The house used to be a fairly standard template but now has several form options. The common thing in all houses is listing the CTQs and seeing the correlation between the CTQs and the technical capabilities of the person working on the project. House of Quality, as the name suggests, is a tool, used in product planning and design, that closely resembles a house.

Today's House of Quality formats vary from complex engineering examples to simplistic graphics. All formats use the CTQ tree (Figure 10.5) in the center of the House of Quality. This tree may be used independently.

DMAIC WBS

A DMAIC WBS may be a useful tool because it provides a map of what will be happening in the future. This is closely related to a project plan. Unlike a project plan, however, the DMAIC model is a discovery model. Therefore, not all steps are included. Also, the DMAIC WBS is a living document and may be continually updated.

Figure 10.5   CTQ tree.

The DMAIC WBS takes each category of the DMAIC and presents it in outline form. The outline shows the anticipated steps in each phase. Even though the DMAIC WBS looks like a project plan, in this case, there is no baseline. It is simply a to-do list using the standard outline, for example,

1. Define
    a. Project charter
    b. Design process map
    c. Perform QFD

2. Measure
    a. Data collection plan
    b. Benchmarking study

The DMAIC WBS is basically a projection of what will happen in each phase. Obviously, the detail won't be available, but there will be a high-level view of the planned accomplishments.

Two popular visual tools used in the Define phase are the affinity diagram and the Kano model or analysis. These tools allow participants to quickly identify key elements of project success. Both tools are useful when a large group of people is working on the same project or when there is conflict within the group.

Affinity Diagram

The affinity diagram (Figure 10.6) organizes a large number of ideas into their natural relationships. This technique works best with medium-sized

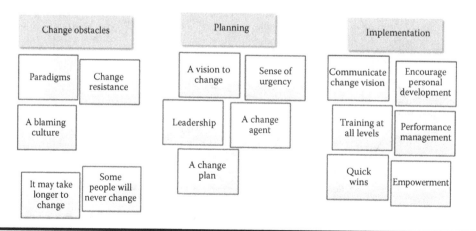

Figure 10.6   Affinity diagram.

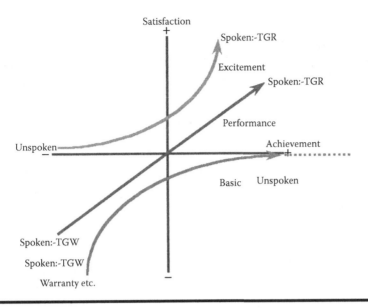

Figure 10.7   Kano diagram.

groups, especially those with diverse viewpoints and team involvement. The first step is simple brainstorming, followed by listing the ideas in categories. Post-It notes are also used to record the group's generated ideas and then to sort them under appropriate headings. It is important that everyone in the group participates and fully understands what each of the ideas means.

Kano Model

A Kano model is a product development theory, now also applied to the service industry, that suggests customer preferences are divided into categories that include delighters, satisfiers, and dissatisfiers. To be successful, a product or service must meet the customers' needs in quality and performance. However, delighters get the customer excited with an unexpected benefit, often resulting in that extra wow factor.

In a Kano model (Figure 10.7), topics are grouped according to how they will delight, simply satisfy, or dissatisfy the customer. Professor Noriaki Kano classified them into these five themes:

Attractive
One-dimensional
Must-be
Indifferent
Reverse

These categories were later reduced to the delighters, satisfiers, and dissatisfiers currently being used.

At the end of each phase in the DMAIC cycle, a tollgate step is used. Think of a tollgate as a checklist of everything that should be done before moving to the next phase. Tollgates will always carry specific components related to the project. In the Define phase, here are some general items that would be included:

CTQs identified and explained
Project charter
Processes mapped
Team readiness

In large companies, it is not uncommon to have a tollgate review at the end of each phase. This review may be done by the PM or a person who is not familiar with the project. People who are not intimate with the project may provide a more objective point of view.

Project management is used throughout all phases of the DMAIC model. However, the most apparent use of project selection and planning tools is in the Define phase. The reason a project manager who is trying to be leaner and more agile should be interested in this chapter is that it explains simple tools that can be used at the ideation stage of a project even if that project is not intended to be a DMAIC one.

Chapter 11

Making the DMAIC Model *Leaner* and More Agile: Measure

Define, *MEASURE*, Analyze, Improve, Control

Edwards Deming, the father of modern quality control, said that anything that is measured gets better. The main purpose of the Measure phase is to establish a clear as-is picture of where the existing process is today and to make sure that the tools used to measure the activity are reliable and valid. In this phase, particular attention needs to be paid to eliminating information that is judgmental or biased. Facts need to be recorded accurately.

In the traditional Define-Measure-Analyze-Improve-Control (DMAIC) model the first step in problem solving is defining the problems and making sure you are working on the correct issues. The second step is getting a clear as-is picture of where you are today. Often project managers don't take the time, or are too busy, to look at the current situation and the impact the project might have on various departments. The value of studying the current state gives project managers insight into issues connected to the project. It also allows the ability to identify possible risks not normally apparent.

When leaving the Define phase, two documents are brought to the Measure phase: the process map and the project charter. The project charter document is used as a reference document throughout the DMAIC phases. The process map is used to decide what functions in the process should be measured. By recording a more comprehensive assessment of the existing process, the detailed process map will help pinpoint the source of the problem.

In the Define phase, the process map is often developed as a high-level, detailed diagram. A detailed process map is a quick way to determine what activities should be measured.

An example of an existing process that might need improvement is the topic of student enrollment. The activities involved in the student enrollment process would be detailed on the current process map. Now as a measurement, how long does it take for each activity to be completed? Can each activity be measured in a reliable and valid way? What specific activities in the student enrollment process are handled by certain individuals? Can this performance or lack thereof be measured? Are there geographic concerns? Are there technical activities that can be measured?

The reliability of the tool being used to measure must be considered in this phase. In manufacturing environments, this may mean the calibration of certain tools. In service industries, in which simple observation is often used to determine how well things are going, it may mean documenting who and how the observations are being made. Is the method used to measure fair?

It is imperative to establish how much time and cost are being used for each activity currently. Without this information, it is difficult to document that the process improvement has been implemented along with its resulting benefits.

In the Measure phase, tasks are related to recording defects, mistakes, or variations and identifying process improvement opportunities. The measurements must be reliable and valid. In Lean Six Sigma, think of reliable as relating to the measurement tool itself. For example, is a measuring tape a reliable instrument to measure inches? Normally, yes. Validity, in this case, means whether it has meaning related to the project. Is the measurement chosen related to the problem that is being solved or the activity being rolled out?

One of the major activities of Lean Six Sigma is gathering data. Other improvement methodologies often attempt process improvement without the appropriate data to understand the underlying causes of the problem. Not having the right data or enough data can result in short-lived or disappointing results.

Key tools include the following:

Process mapping (detailed look)
Input and output definition (Xs and Ys)
Benchmarking
Scorecards
CTQs—measurement
Cp and Pp Index
Failure mode and effects analysis (FMEA)
Sigma calculations
Measurement systems analysis (MSA)
Data collection plan

Detailed Process Map

One of the tasks that should be accomplished before leaving the Define phase is a high-level process map. In the Measure phase, a process map should contain as much detail as possible (see Figure 11.1). Much like a detailed flow diagram, the process map should contain graphic illustrations of all inputs and outputs in the process. A more detailed process map will help to identify relationships to be measured. The process map designed in the Define phase now takes on more detail. The map should be created as a team effort. Anyone who is involved with the process should be involved in creating the map, which will help to obtain buy-in later. One of the goals of the Measure phase is to pinpoint the location or source of a problem as precisely as can be determined. Then measure whatever makes sense to measure to make your product or service better quality, faster, or more cost-effective, which is the objective in Lean Six Sigma.

A key concept in Measure is the X versus Y relationship. X = input. X is also known as key input process variables (KIPVs) or the vital few Xs. Y = output and is also called key output process variables (KOPVs) or the vital few Ys. The $Y = f(x)$ relationship or formula states that what goes into the system (x) is impacted by how x functions (f). The small x in this case may also mean multiple inputs (see Figure 11.2).

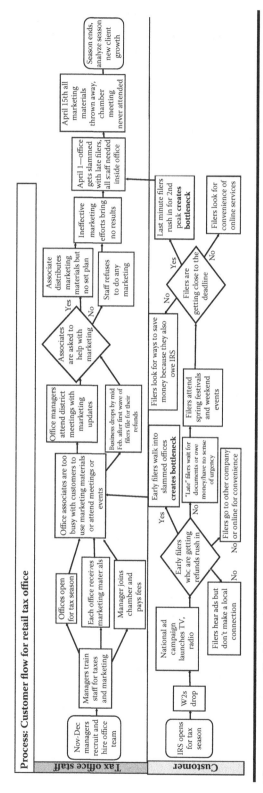

Figure 11.1 Detailed process map measure.

Y = f(x)
Y is the process improvement: how to make it better, faster, or more cost-effective
f is how the xs function
(x) are all the inputs neccessary to create the improvements

How to make better coffee:
Y =   To purchase the best coffee (budget or allowance)
          Given budget, training, competitive pay scale (fresh clean water)
          To keep the cream fresh (refrigerator)
          To grind the coffee beans (coffee grinder)
          To drink the coffee (favorite coffee cup)
          To stir the coffee (utensils)
          To make the coffee and enjoy! (15 minutes)

Internal benchmarks:
YTD % over last year
Positive new client growth measured by a 7-day trend over last year
YTD % higher than the regional average
Other measurements:
Leads converted measured weekly
Leads converted breakdown per activity
Competitive benchmarks:
Pay rate for tax pros and pay rate for marketing
What other tax companies charge per return
The accuracy rate of other tax companies
The marketing experience and promotions from other tax companies

Figure 11.2   Y = f(x) measure.

Benchmarking

Benchmarking is an effective and popular technique to define the as-is state (see Figure 11.3). The benefit of benchmarking is to compare who is performing better and to find the means to improve in those areas of business. There are several different types of benchmarking:

Internal
Competitive
Functional
Collaborative
Generic

Internal benchmarking compares activities and processes within the organization to another operation in the same company. For example, internal benchmarking could be used to share best practices between departments or to increase efficiency. Competitive benchmarking is the most challenging as it compares against competitors, and this information is sometimes difficult to secure. Some industries may have legal constraints

---

Benchmarks for a local bank:

Internal benchmarks (usually easiest to implement):
    YTD % profits over last year
    Profits compared to other branches
    Banking errors compared to last year
Competitive benchmark, comparing a company's position within an industry (often the most difficult):
    Interest rates offered compared to other banks
    Fees that other banks charge
    Convenience of drive-through window
    Branch open hours
Collaborative
    Membership to banking associations
    Safety of deposits and securities
    Works with high schools to establish student savings accounts
Functional
    Compare the application to a loan to the application for insurance
    Compare the process speed for drive-through bank deposit to other drive-throughs
    (pharmacy, burger house, dry cleaner)

Figure 11.3 Benchmarking.

or trade secrets. However, competitive benchmarking could be used in evaluating a competitor's products, total market share, shipping costs, or processing time. When looking at competitors, it is important to focus on what gives your product or service the competitive advantage and to look for innovative ways to sustain it. Functional benchmarking analyzes and compares a function; think of functions as activities. Collaborative benchmarking is the most futuristic as its purpose is to improve industries, such as health care or world initiatives. It is setting the standard for a specific field or industry. Generic benchmarking looks at any activity, operation, or process.

Scorecards

The balanced scorecard (Figure 11.4) developed by Kaplan and Norton at Harvard University in the early 1990s focused on these quadrants: finance, customer service, business process, and learning. A scorecard may pick any four areas related to the project and compile measurements around those categories.

Using scorecards is a good way to gather measurements because each quadrant may already have developed metrics. For example, one common financial measurement would be return on investment.

Another place to start measuring would be the CTQ factors developed in the Define phase. Because the objective is to get a clear as-is picture of

Balanced scorecard Local bank

Vision and strategy

Financial

Goals	Measures
Maximize profitability	Cost to spend ROI
Maximize efficient process of loan applications	Implement new software
Maximize revenue by increasing clients and investments	YTD and 7-day trend NCG prior year
Maximize revenue	1. Target more high-income customers, 2. Offer more incentives to current clients, 3. Add on service features, 4. eliminate misuse or abuse
Maximize cost saving	Cost to spend ROI

Customer

Goals	Measures
Quality of service in office	Honest customer feedback, customer loyalty
Service: accuracy of the account	Competitive rates
FDIC insured	Fast response to loans applications
24/7 phone support	Delivery payment options and advice
Overdraft protection	Choice of savings plans and loans

Internal business process

Goals	Measures
Acquiring new clients	Assessment of current process
Have right employees at the branch	Standardize training, assess communication procedures
Timely database of system updates	Assessment of current interest rates and offers
Leverage national campaign with grassroots marketing activities	Standardize message, increase quality of service
Accurate follow-up of client experience	Standardize marketing materials and offers

Learning and growing

Consumer:	Community:	Bank associates:	Management team:
Free financial-planning seminars	Provides local jobs	Gain experience	Rewarded for high performance
Free savings account for seniors and students	Support nonprofits, sponsorships	Customer feedback available	Works closely with regional team
Good influence for the community	Student financial literacy program advice and support	Free college tuition program	Most up-to-date software provided
	Active in chamber	Great opportunity to move up in company	National management training
	School sponsorships	Choice of convenient locations	Best practices shared among management

Figure 11.4 Balanced scorecard.

where the project is today, determining how well the CTQs are being met is useful.

A prioritization or decision matrix is a useful technique that can be used with team members or users to achieve consensus about an issue. The purpose of the matrix is to evaluate and prioritize a list of options. The team first establishes a list of weighted criteria and then evaluates each option against those criteria.

This tool may be called a Pugh matrix, decision grid, selection matrix or grid, problem matrix, problem selection matrix, opportunity analysis, solution matrix, criteria rating form, or criteria-based matrix.

A basic process for a decision matrix is to make a list of things that are important when choosing a job. Give each factor a rating from 1 to 10. Make a column for each job being considered. If it meets the rating factor, a plus goes in that column; if not, place a minus symbol. The column with the most pluses wins.

Process cycle efficiency is a calculation that relates the amount of value-added time to total cycle time in a process. A Lean process is one in which the value-added time in the process is more than 25% of the total lead time of that process.

There are many presentation templates to show the balanced scorecard results, and often companies have developed their own format. This worksheet may be used to quickly gather the necessary information.

Failure Mode and Effects Analysis

FMEA is a tool that may be used throughout the DMAIC model. In the Measure phase, an FMEA would be used to measure the current process controls. A process control is what is currently in place to manage the risk of something going wrong. FMEAs can also be forms that are used to identify every possible failure mode of a process or product. These forms can also be used to rank and prioritize the possible causes of failure, determine their effects on the other subitems, and develop preventative actions.

Properly used, the FMEA provides several benefits that include the following:

Improving product/process reliability and quality
Increasing customer satisfaction

Early identification and elimination of potential product/process failure
  modes
Documenting risk
Developing actions to reduce risks

FMEA forms may contain a variety of information, and there are many
presentation templates used to record the data. Additional information may
include formulas that calculate risk, occurrence, or detection. Some companies have created their own FMEA forms.

Risk, occurrence, and/or detection may have industry-associated formulas that determine on a scale from 1 to 10 the severity of the problem (risk),
how often the problem may occur (occurrence), or how likely the company
is to catch the problem before it becomes a problem.

In its simplest form, the FMEA is designed to identify the key activities
(functions or processes) in a project and then determine the impact of that
activity not being successful. Other basic factors, such as determining the
reasons the activity could fail and the controls currently in place to avoid
that happening, should be included.

Figure 11.5 shows a worksheet to gather key information that may be
used on any FMEA form or analysis.

Although the basic FMEA information and process is the same in all
FMEAs, it is not uncommon for FMEAs to be applied specifically in one of
the following areas. When this is the case, there is often additional information recorded on the form. Generally, the types of FMEAs include the
following:

System
Service
Software
Design
Process

System FMEAs, for example, may include items such as product specification, design considerations, and company or industry constraints.

A service FMEA may consider additional information specifically related
to one of these areas:

Purchasing
Supplier selection

Measure current process/product
Failure modes and effects analysis (FMEA)
Marketing for a local pizzeria

Process or Product name:	Local marketing for offices			How often does cause and FM occur	Prepared by: Dawna Miller	How well can you detect cause or FM? Page _____ of _____	
Responsible:	Office manager, employees		How severe is the effect to the customer		FMEA Date (Original) _____ 5/14/15 _____ (Rev) _____		Who is responsible for the recom- mended action?
What is the process step	**What is the key process input?**	**In what ways does the key input go wrong?**	**What is the impact on the key output variables (customer requirements) or internal requirements?**	**How severe is the effect to the customer?**	**What causes the key input to go wrong?**	**How often does cause or FM occur?**	**What are the existing controls and procedures (inspection and test) that prevent either the cause or the failure mode? Should include an SOP number.**	**How well can you detect cause or FM?**	**What are the actions for reducing the occurrence of the cause, or improving detection? Should have actions only on high RPN's or easy fixes.**
Local marketing for pizza shop	Getting clients to come in to buy pizza	Customers don't come in buy to pizza	New client growth and revenue is declining. 1# After four years company could possibly go out of business	Severe	No strategic marketing plan, no standardized training, short staffed	80%	Pizza managers attend weekly regional meetings for marketing updates, specials, materials to be distributed locally.	Not well	Find another way to get the message out to potential clients	Pizza manager, employees
Employees are respon- sible for marketing	Employees distribute flyers and local marketing between busy shifts	Employees do not go outside of pizza restaurant 1# to market	Lost opportunity and wasted marketing hours/ materials	Moderate	Employees are too busy delivering hot pizza or they do not want to leave the office and hate marketing	90%	"Marketing" is listed under the job description. Each employee is to spend 3 hours a week giving out flyers and attending events. Nothing is done for marketing because there is no accountability.	Not well. Can only be checked by having employees take photos of activities	Scorecard filled with ratings on office standards including a marketing tracker filled out and signed by manager	Upper management conducts spot check by surprise office visits
Competi- tor's come out with new ad campaigns and huge discounts	Pricing—try to match competitors offers	Cannot match competitor ads or discounts	Customer goes to competitor	Severe	Company does not respond quick enough to match the discount, company will not match the discount, 1# Discount is not well understood by the customer	50%	Updates from the national media department, word of mouth including customers. 1# Controls include national marketing team's ability to act quickly and mobilize a marketing street team.	Well	Better communication by sending the updates directly to all employees, and posting the notices on the wall	National marketing department. Occasionally the managers can offer discounts from the office level

Figure 11.5   FMEA worksheet.

Payroll
Supplier payment
Customer service
Recruitment
Sales
Logistics
Project planning
Scheduling of services

Software FMEAs may include various system analytics or industry benchmarks whereas design and process FMEAs may include customer requirements and standards.

The following section is for project managers who want to have more of an advanced knowledge of measure. These concepts are not necessary to run a successful Lean Six Sigma project.

Sigma Calculations

Sigma calculations are very useful in the Measure phase to determine the current sigma level of an activity or transaction. In order to calculate the defects-per-million-opportunities (DPMO), three distinct pieces of information are required:

The number of units produced
The number of defect opportunities per unit
The number of defects

The actual formula is

$$DPMO = (\text{number of defects} \times 1{,}000{,}000)$$

$$(\text{Number of defect opportunities/unit}) \times \text{number of units}$$

To use the sigma level effectively as a form of measurement, two conditions must exist. First, the item must be something that can be counted. Second, everyone must agree on what constitutes a mistake or defect.

Cp and Pp Indexes

Several statistics may be used to measure process capability. A capable process is one in which almost all the measurements fall inside the specification limits. A measurement similar to sigma is capability metrics (Cp), which measures the process capability. A technique used to determine how well a process meets a set of specification limits is called a process capability analysis. A capability analysis is based on a sample of data taken from a process and usually produces the following:

An estimate of the DPMO
One or more capability indices
An estimate of the sigma quality level at which the process operates

A process is capable if it falls within the specification limits. Graphically, the process capability is accomplished by plotting the process specification limits on a histogram or control chart. If the histogram data fall within the specification limits, then the process is capable. Often manufacturing environments prefer to use Cp. Traditionally, if Cp is measured at 1 or higher, the index is indicating that the process is capable. In manufacturing, the number often needs to be at 1.33, which is the same as 4 sigma. The number 2 in the index represents 6 sigma. The process capability index, or Cpk, measures a process's ability to create a product within the specification limits.

Process capability refers to the ability of a process to produce a defect-free product or service in a controlled manner. This is often measured by an index. The process capability index is used to find out how well the process is centered within the specification limits.

A more sophisticated capability analysis is a graphical or statistical tool that visually or mathematically compares actual process performance to the performance standards established by the customer. In summary, the capability of processes may use indices called Cp and Cpk. These two indices, used together, can tell us how capable our process is and whether or not we have a centering issue. Cp is the potential capability of the process; CpK is the actual capability of the process.

Because a histogram is used to track frequency, a quick way to study process capability is to review the size and shape of the histogram. For example, if the histogram's bar shape looks like a normal distribution (most things are centered in the middle and it has a bell shape), the process would appear to be capable of handling the majority of issues. If, on the other hand, the bell is leaning to the right or left, this may indicate an opportunity for process improvement.

A nonmathematical way to determine capability is to thoroughly examine the CTQ objectives. While the production is in progress, the performance of the process is monitored to detect and prevent possible variations. A process is considered capable if the process mean is centered to the specified target and the range of the specified limits is wider than the one of the actual process variations.

As mentioned in the Define phase, each phase is completed by reviewing a tollgate. Once again, there will always be items specific to the project that will be included in the tollgate. At this tollgate, a major item would be the data collection plan—ensuring that it has been established and documented and that data have been collected on key measurements. Remember that the objective is developing a clear picture of the current process.

A basic understanding of financial management is useful in the Measure phase. Just as Lean Six Sigma assumes there is a basic knowledge of project management, there is also an assumption that practitioners have been exposed to information such as finance and accounting for the non-financial manager. This is necessary because many measurements may be based on finance or basic accounting.

Although no one expects an ILSS practitioner to have the same financial knowledge as a CPA or CFO of a company, the following concepts are essential and not covered in this material:

Accounting terminology and underlying concepts
The role of various financial statements
Distinguishing income from cash flow
The accounting processes
The quality of earnings
Financial decision making
Analysis of financial reports
Approaches to valuation
Calculating return on investment

Regardless of how information has been collected, one of the key factors in the Measure phase is the question of reliability and validity.

Reliability is the consistency or stability of indicators. A reliable instrument yields the same results on repeated measures. An instrument may be reliable but not valid.

There are different types of validity, including:

Face validity (assumptions of a logical tie between the items of an instrument and its purpose)

Content validity (the items in the instrument are systematically judged by a panel of experts and rated as to the extent to which the item adequately represents the construct proposed)

Criterion-related validity (what is the relationship between the subject's performance on the measurement tool and the subject's actual behavior?)

Measurement Systems Analysis

MSA is a mathematical method of determining how much the variation within the measurement process contributes to the overall process variability. It takes into consideration bias, linearity, stability, repeatability, and reproducibility.

MSA actually builds a foundation in the Measure phase. It is a component of analysis as to how the data were gathered. The reliability of the data may be questioned if the team presents results that are not favorable.

An MSA is a specially designed experiment that seeks to identify the components of variation in the measurement. Just as processes that produce a product may vary, the process of obtaining measurements and data may have variation and produce defects.

MSA is a process that was developed by the nonprofit organization Automotive Industry Action Group (AIAG).

AIAG concentrates on the following measurement analysis conditions:

Bias
Stability
Linearity
Repeatability and reproducibility

Bias is a measure of the distance between the average value of the measurements and the true and actual value of the sample or part.

Stability refers to the capacity of a measurement system to produce the same values over time when measuring the same sample.

Linearity is a measure of the consistency of bias over the range of the measurement device.

Reproducibility assesses whether different appraisers can measure the same part or sample with the same measurement device and get the same value.

Repeatability assesses whether the same appraiser can measure the same part or sample multiple times with the same measurement device and get the same value.

Data Collection Plan

A useful tool in the Measure phase is a data collection plan. The purpose of a data collection plan is twofold. First, it is a communication tool. Second, it is a strategy for how the right data will be collected and used. The choice and logic of the measurement must also be articulated. For example, if sigma will be used to measure the project, a short description of sigma should be included.

In many cases, this can be accomplished by examining the data itself to check for items that seem to be out of order. A data collection plan should document the phases, including pre-data collection and post-data collection. The five steps involved should include the following:

Clearly defining the goals
Reaching an understanding (authority)
Ensuring the data are reliable
Collecting the data
Following through with results

In order to move to the next phase, measurements of the key aspects of the current process must to be completed along with the collection of all the relevant data. During the Measure phase, the focus is on gathering data to describe the current situation.

Data collection can also be accomplished by placing data in a Pareto or histogram chart. These charts are more fully discussed in the Analyze phase. Histograms are used to track the frequency of events whereas Pareto charts are used to track the types of events. For example, if the number of calls into a help desk were being tracked, a histogram would be used. If the types of calls were being tracked, a Pareto chart would be appropriate. Both charts are usually represented as bars. The Pareto chart has the bars represented from largest to smallest. The histogram represents the frequency as the information is collected. In the Define phase, a Pareto chart may be used as part of the historical data to determine the largest opportunity or threat. In the Measure phase, the chart would be used strictly for data collection purposes. This information would be further studied in the Analyze phase.

Remember that when leaving the Measure phase of the DMAIC model, the goal is to be able to present a clear as-is picture of the various activities in the process. How much are these activities costing now? How long do these activities take now? Who or what department is responsible now? Are there any bottlenecks occurring in the process now?

The information in this chapter provides tools to quickly answer these questions. The basic concepts provided are important to project managers regardless if they intend to use the DMAIC model.

Chapter 12

Making the DMAIC Model *Leaner* and More Agile: Analyze

Define, Measure, *ANALYZE*, Improve, Control

Most project managers will be aware of some of the tools Analyze uses in the Define, Measure, Analyze, Improve, Control (DMAIC) model because they are popular in many different bodies of knowledge. However, *Leaner*™, discussed in previous chapters, simplified the dynamics of the tools and made them easier to understand. Regardless if a project manager plans to use the DMAIC model or not, certainly a certain amount of analysis becomes part of any project.

Once it is determined that the data received from the Measure phase are both reliable and valid, it is time for the Analyze phase to begin. Remember, the reason data are being analyzed is to find three to five possible solutions, and sometimes the best explanation is the simplest. To find these solutions the following factors should be considered as the data are being analyzed:

Correlation: Is there a common bond (positive or negative)?
Root cause: Why are things the way they are?
What are areas that can be stabilized (variations)?
What impact do the variations have on the solution?

In Lean Six Sigma, creativity is encouraged when analyzing data, keeping in mind that "the voices" of the customer, the business, the employee, and the process are valuable when making decisions. This is done by asking the right questions and using the right tools to identify, clarify, and reveal ideas.

A number of tools are available to analyze data. In fact, many Lean Six Sigma classes focus on the tools alone. Remember, with Lean Six Sigma, start with the easiest tool first. If that tool doesn't yield the results, try something more sophisticated. It isn't always necessary to select a complicated tool. By using this approach, by the time a more sophisticated tool is needed, all the necessary data will have been collected.

When entering the Analyze phase of the DMAIC model, the project manager (PM) now has a clear idea of what process improvement needs to be explored (Define) and the current picture (Measure). These two crucial pieces often lead to immediate ideas on how the process may be improved. Now, however, the data and information gathered in the Define and Measure phases must be analyzed. Sometimes analyzing the information simply involves putting the data in a chart or graph to make them easier to digest. Sometimes, depending on the complexity of the process improvement, it requires more thought and/or more complex data analysis.

The objective of the Analyze phase is to leave with three to five solid process improvement solutions. Each solution needs to be tested for a variety of conditions. These conditions include, but are not limited to, concerns regarding affordability, sustainability, time frames, and capability issues.

This requires not only listing all the possible solutions to the problem, but also running what-if calculations on these solutions. Only those solutions that meet the conditions of the what-if calculations would make the final list. This list is then passed on to the Improve phase.

The Analyze phase is often the most labor-intensive. In fact, many formal training classes spend the bulk of the classroom instruction working with the tools that are used in this phase. Many of the tools reviewed in Define or Measure are also used in the Analyze phase but may take on a different aspect. One of the objectives in this phase is to identify and analyze the gaps between current performance and desired performance.

Additional activities in very advanced projects may also include the following:

Identifying variation
Determining a vital few Xs, $Y = f(x)$ relationship
Determining root cause(s)
Evaluating impact

Because Lean Six Sigma is concerned with both reducing waste and eliminating defects, identifying the sources of the variation is key. Variation is the fluctuation in process output: an occurrence of change or a magnitude of change. Identifying the variation is often done in the Measure phase. In the Analyze phase, although additional variation will become apparent, the main question is what does the variation mean and how does it impact the project?

Whereas, in the Measure phase, the Ys and Xs may have been identified, the Analyze phase concentrates on which are the most important Ys and Xs. Sometimes this will be referred to as the vital few Xs and Ys. They are also referred to as key process input or output variables (KPIVs or KPOVs).

Determining the root cause can be as simple as using a tool known as Five Whys or using a more sophisticated failure mode and effects analysis (FMEA), which were discussed in the Measure phase.

In large projects, it becomes necessary to determine what the vital inputs and outcomes are as opposed to all of the inputs and outputs, so the project manager can focus on which problem is causing the most issues.

In Analyze, the objective is to determine the causes of the problems and decide which specific issues need improvement. Designing strategies to eliminate the gap between existing performance and the desired level of performance is often part of this phase. This involves discovering why defects are generated. This is done by identifying the key variables that are the most likely suspects in creating variation.

Key Tools: The Big Seven

The tools that are primarily used in the Analyze phase are known by a number of different names.

General:

Flowchart/process mapping
Pareto chart
Histogram
Scatter diagram
Fishbone
Check sheet
Control chart

The most important thing to remember about the above tools is that they are designed to help capture a massive amount of data that might otherwise be hard to digest. So, in other words, if there are not a lot of data points or if the answer is obvious, a chart or graph may not be necessary. Another important point is that these are critical thinking tools. Whereas they are more often used to report a condition or in a presentation, their true value is their ability to analyze the data more effectively.

Flowchart

Flowcharts or process maps have been discussed. However, there are a number of flowchart types that can be used to map a process, such as swim lane charts, value stream maps, and spaghetti diagrams. An example of a simple traditional flowchart using flowcharting symbols showing the customer service process is shown in Figure 12.1.

There are a number of other flowcharts that can be used. The purpose of a flowchart is to show a picture or a snapshot of the process or project. An excellent way to get an idea of the physical layout of a process is called a spaghetti diagram.

A spaghetti diagram is a visual representation using a continuous flow line tracing the path of an item or activity through a process. The continuous flow line enables process teams to identify redundancies in the workflow and opportunities to expedite process flow.

This is an excellent example offered by the American Society for Quality (ASQ). ASQ is a global community of people dedicated to quality who share the ideas and tools that make our world work better. With individual and organizational members around the world, ASQ has the reputation and reach to bring together the diverse quality champions who are transforming the world's corporations, organizations, and communities to meet tomorrow's critical challenges. Figure 12.2 shows an example of an employee's day. In this case, the diagram is used to identify potential wasted steps or time in the day.

A swim lane is a visual element used in process flow diagrams or flowcharts that visually distinguishes job sharing and responsibilities for subprocesses of a business process. Swim lanes may be arranged either horizontally or vertically.

The lanes or columns can be named after entries, such as individual names, departments, functions, and times of the year. An example can be seen in Figure 12.3. This is also called a Swimlane Chart.

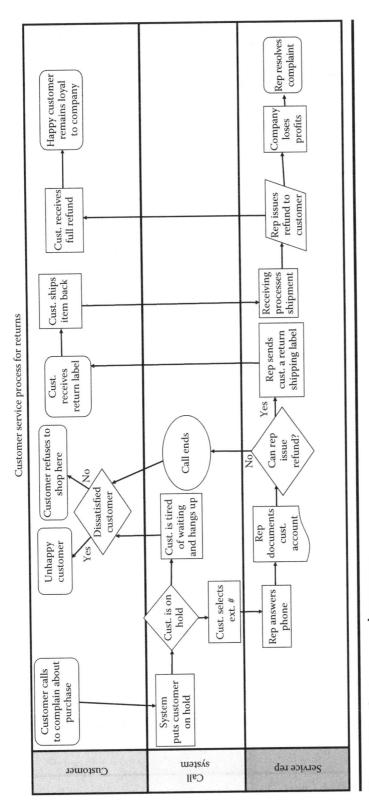

Figure 12.1 Customer service process.

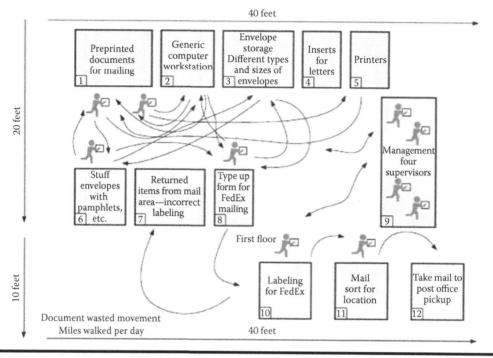

Figure 12.2 Physical movement in an employee's day.

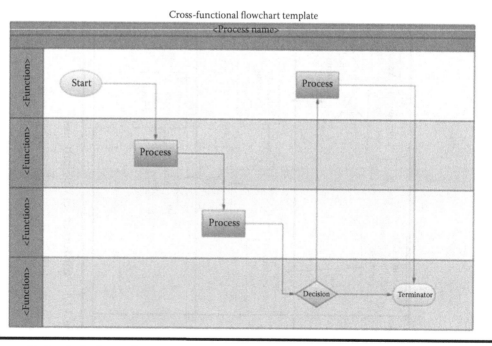

Figure 12.3 Swimlane Chart.

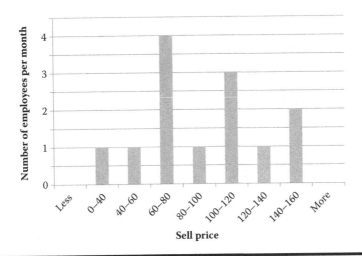

Figure 12.4   Example of a Histogram.

Histogram

A histogram is a graphical representation of the distribution of numerical data and for showing the shape of distribution. A histogram is the most commonly used graph to show frequency distributions.

Histograms are useful when the data analyzed are numerical or when determining the output of the process distribution. This can be useful when analyzing customer requirements or a supplier's process. It is also a quick way to discover if a process has changed from one time period to another (see Figure 12.4).

Pareto Chart

A Pareto chart, named after Vilfredo Pareto, is a type of chart that contains both bars and a line graph; individual values are represented in descending order by bars, and the cumulative total is represented by the line. Pareto charts are useful when analyzing data about the frequency of problems. Whereas a histogram may show you how often a problem occurs, a Pareto chart will show the reasons for the problem (see Figure 12.5 for two examples).

Scatter Diagrams

A scatter diagram, also called a scatterplot, is a visualization of the relationship between two variables. The scatterplot can give you a clue that two

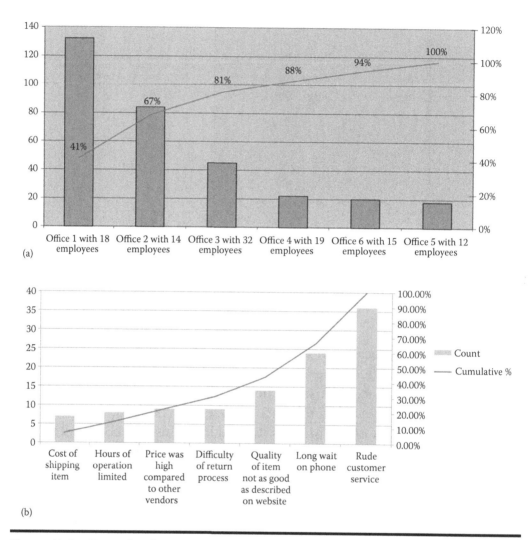

(a)

(b)

Figure 12.5   Example of Two Pareto Charts.

things might be related. Examples could be as simple as temperature versus ice cream sales or complex as in cost versus valve production in a nuclear plant. A simple example is shown in Figure 12.6.

Fishbone Diagram

A very effective but underutilized tool when trying to determine root cause is called the fishbone diagram. This graph is also known as a cause-and-effect diagram or an Ishikawa diagram. It is in the shape of a fish but can take other forms and still be effective. The problem or opportunity is placed

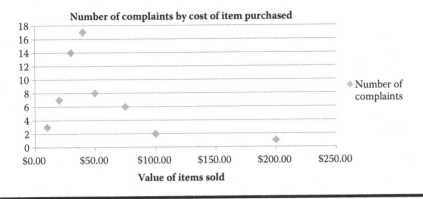

Figure 12.6 Example of a Scatter Diagram.

in the head of the fish, and the fish bones represent contributing factors. The major bones (factors) can then result in smaller bones. The further a bone results in smaller bones, the more likely it is that a root cause can be discovered that would otherwise not be obvious. But sometimes the major contributors (large bones) are all that are needed to discover the root cause. The biggest complaint about fishbone diagrams is they don't solve problems. This is true as the purpose is only root cause identification (see Figure 12.7).

Check Sheet

The easiest way to think of a check sheet is as a simple to-do list. Things are written on the list, things that need to be accomplished and checked off when they are done. But a check sheet can also be used to gather

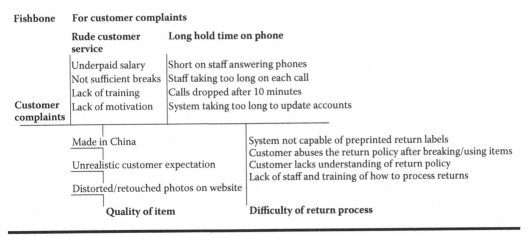

Figure 12.7 Example of Fishbone logic.

Telephone interruptions

Reason	Day					
	Mon	Tues	Wed	Thurs	Fri	Total
Wrong number	ⱵⱵ	‖	‖	ⱵⱵ	ⱵⱵ ‖	20
Info request	‖	‖	‖	‖	‖	10
Boss	ⱵⱵ	‖	ⱵⱵ ‖	‖	‖‖‖‖	19
Total	12	6	10	8	13	49

Figure 12.8 Example of a Check Sheet.

information for any of the other tools, which is why this tool always makes the top-seven list. Figure 12.8 shows a simple example of a check sheet.

Control Charts

To explain a control chart, it is helpful to know about two other charts first: the line chart and the run chart.

A line chart or line graph is a type of chart that displays information as a series of data points called markers connected by straight line segments. It is a basic type of chart common in many fields. When working with a line chart, the designer does not necessarily need to know why he or she is studying the condition, just that it is a condition worth tracking. The chart should be simple and to the point. The Y axis, which is located to the left, and the X axis located at the bottom of the graph need to have clear and concise names to make the line chart easy to understand (see Figure 12.9a).

The run chart has one additional piece of information that the line chart does not have, and this is a middle line that shows the mean average or the median of all the data points. First, all the data points are collected. The mean or median is then given a value, and a line is drawn in the middle of the graph. The Y-axis values are then determined by the middle line. So, if the middle line is 83, the tick mark below 83 would be 82, the line above 84. But if the middle line is 85, the choice could be made that the tick mark above is 90 and the tick mark below is 80. When working with large figures, such as millions of parts or dollars, these tick marks become more valuable. An example of a run chart is shown in Figure 12.9b.

The control chart adds one final piece of information to the chart, and that is the upper and lower control limits (UCL and LCL). These may also be upper and lower specifications or upper and lower tolerance levels.

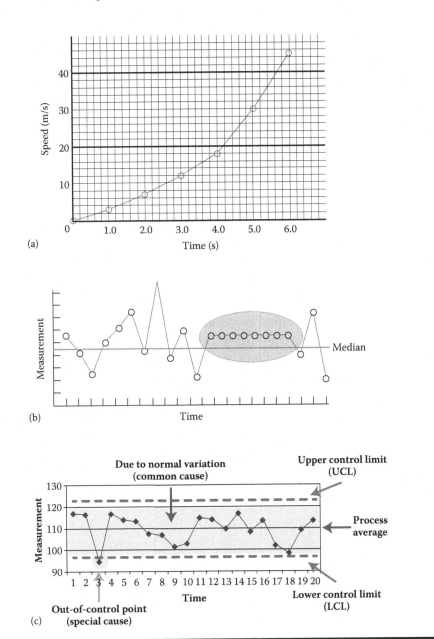

Figure 12.9   Control Chart logic: a) line chart, b) run chart, and c) control chart.

Control charts show items that could be considered outliers and need to be addressed. The look and feel of a control chart are the same, but the interpretation of what the data mean is different. An example is Figure 12.9c, in which items are noted that could be common cause variations. Common cause variation is expected variation or variation by default. For example, if a patron purchases a car, the car will eventually need new tires. Special cause variations are things not typically within our control, such as bad weather or a change in leadership.

The control chart is a graph used to study how a process changes over time. Data are plotted in time order. A control chart always has a central line for the average, an upper line for the UCL, and a lower line for the LCL. These lines are determined from historical data.

A control chart would have these components:

A clear title
Labels on the Y and X axes
Appropriate scale
A middle line, typically indicating a mean (but it could have other meanings)
UCL and LCL

The idea behind statistical quality control is simply to make sure those goods and services satisfy the customer's needs.

The Analyze phase of the DMAIC model uses the information gleaned from the seven tools of process improvement, quality, and critical thinking in a unique way. For example, a process map is used to visually show the steps in a process. In Define, it was used as an information tool and in Measure as a way to decide what measurements should be taken; in Analyze, it would be used to determine bottlenecks and process flow. It is often used as a pencil-and-paper tool with which steps in the process are moved around and/or eliminated to see the total impact on the process improvement.

A fishbone analysis in Define may be used to determine if the right problem is being explored, and in the Measure phase, a fishbone may be used to determine variations that should be measured. In Analyze, the fishbone is almost always used to discover the root cause.

Histograms and Pareto charts used in the Define phase determine if a problem exists. In Measure, these tools are used to gather information on the current as-is state, and in the Analyze phase, they are used to place information in bins or buckets to determine root cause.

A check list is used to collect information in all phases of the DMAIC model. Scatter diagrams are generally specific to the Analyze phase because they are used to show correlation, and control charts are used to analyze patterns within the process. Control charts are often used in the Measure phase to collect current as-is data and in the last phase, Control, as a sustainability tool.

The Five Whys

Although the Five Whys tool is not part of the seven tools of quality, this process is often considered the eighth tool in the most important tools.

The Five Whys is a simple but effective analysis tool to determine root causes. A question is asked five times based on information received in the previous answer until a conclusion is reached. Sometimes this simple tool can solve the problem. It can also determine relationships between root causes of problems and be learned quite easily, and it requires no statistical analysis.

In Analyze, there are also more sophisticated tools that are used for mega-data, extremely difficult solutions, or safety issues. Generally speaking, statistical software is needed to use these tools appropriately, although several can be reasoned out manually in a spreadsheet, such as a correlation analysis. A correlation analysis determines if two variables have a positive relationship, a negative relationship, or no relationship at all. This analysis can be done using a simple scatter diagram. Many times, it is helpful to think of a correlation in the terms of one thing versus another. For example: eating soup versus snowy days or rainy days versus employees who are late to work. A regression analysis is similar to a correlation analysis. Although correlation analysis assumes no causal relationship between variables, regression analysis assumes that one variable is dependent upon the other. Both correlation and regression analyses use scatter diagrams, discussed earlier in this chapter, to represent relationships.

Statistical Thinking

In a Lean Six Sigma project, much of the Analyze phase may be accomplished by using less complicated tools, such as Pareto charts or histograms. But the Lean and Agile PM should have a high-level idea of the following concepts in case the project or process improvement becomes complicated. Knowing that there are avenues available to reason out a problem is helpful regardless if this piece needs to be outsourced or not.

These are topics that usually require hands-on experience in a real project to totally grasp the concept. Various software packages are available to make the process easier. These concepts include the following:

Statistical hypothesis testing
Statistical analysis (statistical process control)

A hypothesis is a tentative statement that proposes a possible explanation to some phenomenon or event. A useful hypothesis is a testable statement.

Usually, a hypothesis is based on some previous observation, such as noticing that in the winter many trees lose leaves and the weather is colder. Are these two events connected? How are they related? The null hypothesis statement would be many trees lose their leaves in winter due to the cold weather. The alternative hypothesis would be that many trees lose leaves in winter, but it is not due to the cold weather.

Sophisticated tools used for hypothesis testing are performed via statistical software. They are only necessary if there are a great deal of data to digest. To conduct a hypothesis test, it is first necessary to determine the premise. This premise becomes the alternative hypothesis, and the opposite statement becomes the null hypothesis. The null hypothesis is mutually exclusive, which means if the alternative hypothesis is true then the null hypothesis is untrue. In theory, hypothesis testing is a great way to analyze data if there are a lot of data.

A popular tool for hypothesis testing is analysis of variance (ANOVA). An ANOVA is an analysis of the variation present in an experiment. Another tool in hypothesis testing is probability models. Using basic probability theory, certain tests are applied to determine the likelihood of something happening.

A statistical hypothesis test is a method of making statistical decisions from and about experimental data. Null hypothesis testing just answers the question of how well the findings fit the possibility that chance factors alone might be responsible. Fortunately, this analysis is done by using statistical software. Generally speaking, it is only necessary when working with extremely large pieces of data.

Hypothesis testing is used to formulate a test regarding a theory that is believed to be true. A common example would be claiming that a new drug is better than the current drug for treatment of the same symptoms. The null hypothesis, H0, represents a theory that has been put forward as true. The alternative hypothesis, H1, is a statement of what a statistical hypothesis test is set up to establish. In a hypothesis test, a type I error occurs when the null hypothesis is rejected when it is in fact true, that is, H0 is wrongly rejected. A type II error occurs when the null hypothesis, H0, is not rejected when it is in fact false. When working with hypothesis testing, terms that are likely to be used are the following:

p value
t test
ANOVA

A p value, a hypothesis testing component, is the probability value (p value) of a statistical hypothesis test: the probability of getting a value of the test statistic better than that observed by chance alone.

The t test assesses whether the means of two groups are statistically different from each other.

An ANOVA is a mathematical process for separating the variability of a group of observations into assignable causes and setting up various significance tests. This is a statistical technique performed in a statistical package designed to analyze experimental data. Minitab Statistical Software is the leading statistical package used to analyze data for Six Sigma, but a number of other packages exist. Most statistical analysis for Lean Six Sigma may be performed using MS-Excel; however, some of the more sophisticated exercises, such as ANOVA, are better performed using statistical software.

An ANOVA may be used even when hypothesis testing is not being considered just as a general information tool. Whereas a t test only looks for the difference in mean (average) between two data sets and an F test looks at more than two data sets to determine the mean, an ANOVA also calculates items such as median, mode, maximum, minimum, and confidence level.

Statistical Process Control

Statistical process control (SPC) is building and interpreting control charts. A center line—usually the mean—is established. Then UCL and LCL lines are added. If the data points fall between the UCL and LCL, the process is considered to be in control or stable.

Statistical analysis is often called SPC. The primary tool used in SPC is the control chart. SPC involves using statistical techniques to measure and analyze the variation in processes. Most often used for manufacturing processes, the intent of SPC is to monitor product quality and maintain processes to fixed targets.

In a control chart, discussed earlier in this chapter, the main objective is to keep the process stable, which means that the process functions within the UCL and LCL, specification limits, or tolerance limits. The science of SPC believes that how the process reacts within those specific limits also provides valuable information.

As a reminder, a run chart simply collects data points and charts them on a graph. A control chart, on the other hand, establishes a mean line,

UCL, and LCL. The purpose of a control chart is to determine if a process is stable. A process is stable if the data points fall within the UCL and LCL.

There are a number of charts that may be utilized to give specific SPC information. A brief description of each chart is given.

Control charts fall into two categories: variable and attribute control charts. Variable data can be measured on a continuous scale, such as a weighing machine or thermometer. Attribute data are counted, for example, good or not good, true or false, broken or not broken. Some control charts are better for illustrating variable data, and some are better for depicting attribute data.

There are many different charts, but some of the most popular are the following:

X-bar/R chart
P chart
Np chart
C chart
U chart

The X-bar/R chart is normally used for numerical data that are captured in subgroups in some logical manner, for example, three production parts measured every hour. A special cause, such as a broken tool, will then show up as an abnormal pattern of points on the chart. It is really two charts: an X-bar and a range chart. The X-bar chart monitors the process location over time based on the average of a series of observations, called a subgroup. The range chart monitors the variation between observations in the subgroup over time.

A P chart is an attributes control chart used with data collected in subgroups of varying sizes. Because the subgroup size can vary, it shows a percentage on nonconforming items rather than the actual count. P charts show how the process changes over time. The process attribute is described in a yes/no, pass/fail, go/no-go form.

Np charts also show how the process, measured by the number of nonconforming items it produces, changes over time. The process attribute (or characteristic) is always described in a yes/no, pass/fail, go/no-go form. For example, the number of incomplete accident reports in a constant daily sample of five would be plotted on an Np chart. Np charts are used to determine if the process is stable and predictable as well as to monitor the effects of process improvement theories.

The C chart evaluates process stability when there can be more than one defect per unit. The C chart is useful when it's easy to count the number of defects and the sample size is always the same. It is often referred to as simply the count.

A U chart is an attributes control chart used with data collected in subgroups of varying sizes. In U charts, it is shown how the process measured by the number of nonconformities per item or group of items changes over time. Nonconformities are defects or occurrences found in the sampled subgroup. They can be described as any characteristic that is present but should not be or any characteristic that is not present but should be. For example, a scratch, dent, bubble, blemish, missing button, and a tear are all nonconformities. U charts are used to determine if the process is stable and predictable as well as to monitor the effects of process improvement theories. The U chart is used to count things by units. Sample sizes may be constant or variable.

Statistics and business/financial math can be used in any phase of the DMAIC model, but the Analyze phase is an appropriate place to do a primer because analyzing data often depends on these two sciences.

First, consider the order of operators when looking at any formula. This is a topic that comes up constantly in all of our mathematical work. The order is parenthesis, exponents, multiplication, division, addition, and finally subtraction (PEMDAS). Avoid the tendency to do the multiplication first.

Mean, median, and mode are three kinds of ways to measure the middle. In Lean Six Sigma, because the goal is to standardize, which means bringing things to the middle value, having three different measures of central tendency can be helpful. If the mean, mode, and median are drastically different, it may indicate that we need to reassess where we think the middle falls.

The mean, or average, of a set of numbers is found by dividing the sum of the numbers by the amount of numbers added.

Example: What is the mean of these numbers?

6, 11, 7

Add the numbers: 6 + 11 + 7 = 24.
Divide by how many numbers (there are three numbers): 24/3 = 8.

The mean is 8.

The median is the number in the middle. If there is an even set of numbers, the two middle numbers are divided.

Example: Find the median of 12, 3, and 5.
Put them in order:

$$3, 5, 12$$

The middle is 5, so the median is 5.
The mode is the number that appears most frequently.

Example: What is the mode?

$$3, 7, 5, 13, 20, 23, 39, 23, 40, 23, 14, 12, 56, 23, 29$$

In order, these numbers are

$$3, 5, 7, 12, 13, 14, 20, \mathbf{23}, \mathbf{23}, \mathbf{23}, \mathbf{23}, 29, 39, 40, 56$$

This makes it easy to see which numbers appear most often.

In this case the mode is **23**.

The only tricky thing with these measures of central tendency is that with medians the numbers must first be placed in sequential order. In all three measures, remember that if a certain number appears multiple times, it has to be recorded each time.

Range is the highest number in a data set minus the lowest number in the data set. How many numbers there are in the data set often determines how confident we are that we gathered the right amount of data.

Stem-and-Leaf Diagram

One simple way to view the data that have been collected is by using a stem-and-leaf diagram, and another would be to use a frequency table. Either way, once data are collected, they must be organized in a logical way so that the viewer can draw appropriate conclusions about what the data represent.

In a stem-and-leaf diagram, the first number becomes the stem, and any numbers after that become the leaves. So, for example, if the number set was 12, 13, 23, 24, 26, 31, 32, and 33, the diagram would look like this:

1: 2, 3
2: 3, 4, 6
3: 1, 2, 3

A frequency diagram is a way of tabulating data in which the independent variable is listed in the left-hand column. The frequency, which is the number of times the independent variable occurs, goes in the right-hand column. If we take this information and make a bar chart, it can also be known as a histogram.

Example: Frequency diagram
Sarah did volunteer work:

Saturday morning
Saturday afternoon
Thursday afternoon

The frequency was 2 on Saturday, 1 on Thursday, and 3 for the whole week.

With both a stem-and-leaf diagram and frequency diagram, we are trying to determine how often things occur. Sometimes a frequency diagram will be expanded to show relevant frequency. In other words, a third column may be added to explain how much of the percentage the sample represents of the overall population.

Type I and Type II Errors

How a sample is drawn from a population is critical to analyzing data. A population is a collection of data whose properties are analyzed. The population is the complete collection to be studied; it contains all subjects of interest. A sample is a part of the population of interest, a sub-collection selected from a population. There are many factors to consider when choosing a sample. Lean Six Sigma is generally concerned with the size. The larger the sample, the less likely a mistake will be made.

A larger sample contributes to avoiding Type I or Type II errors, which are simply a false positive or a false negative. The reason a sample is being studied is that it would not be practical to study the entire population before making a claim. If the whole population was studied and an assertion was made, it would have close to 100% confidence level that a Type I or Type II error has not happened. Anything less than 100% lowers the confidence level in the data.

Often the decision to study only a sample is based on time, resources, and availability of data. Ten percent of the population for a sample is a good rule if the population isn't very large or isn't too small. For example: If ten surveys were sent out and only 10% were returned, one survey, this would not be enough data to make an inference or draw a conclusion. Likewise, if a million products were produced by a rather small company, choosing 10% or 100,000 to study may not be realistic as far as time, money, and resources are concerned.

Design of Experiment

Statistical data and statistical data packages are often used to help in the decision making if there is a large amount of data. Data that can be handled on a simple spreadsheet do not typically need sophisticated tools. Because most spreadsheet packages handle pivot tables, which do a good job filtering data, sometimes tools, such as ANOVAs and design of experiments (DOEs), are unnecessary.

MS Excel pivot tables are the easiest for filtering data. By using a pivot table, summary information may be summarized without writing a single formula or copying a single cell. The most notable feature of pivot tables is that data are arranged in a logical order. Creating neat, informative summaries out of huge lists of raw data is valuable when digesting the data.

When there are a number of variables and those variables have a number of characteristics, in order to analyze data, it might be necessary to use a DOE. A DOE is a tool available in most statistical packages. The term experiment is defined as a systematic procedure carried out under controlled conditions. DOEs, or experimental design, are the design of all information-gathering exercises in which variation is present, whether under the full control of the experimenter or not.

The characteristics of a DOE include the following:

Planned testing.
Data analysis approach is determined before the test.
Factors are varied simultaneously, not one at a time.
Very scientific approach.

DOEs are also powerful tools to achieve manufacturing cost savings by minimizing process variation and reducing rework, scrap, and the need for inspection. Because these designs have become very sophisticated, the typical role of the ILSS practitioner is not to create the design, but rather make sure the appropriate, most reliable, and valid data are entered into the spreadsheet before the DOE is initiated.

Much of the DOE information can be gathered via a histogram. Other components that may also be used as stand-alone tools include SPC and regression analysis.

In SPC, the PM makes interpretations primarily from control charts. SPC is basically a decision-making tool. When a process goes beyond the agreed-upon limits, control charts can help the PM determine the appropriate change. See control charts earlier in this chapter.

Analysis of Variance

The basic ANOVA compares the means of two different groups, which is also known as a t test. ANOVAs have more sophisticated functions and can perform things such as identifying the possibility of a Type I or Type II error. There are several types of ANOVAs depending on the number of treatments and the way they are applied to the subjects in the experiment.

One-way ANOVA is used to test for differences among two or more independent groups.

A two-way ANOVA is used when the data are subjected to repeated measures, in which the same subjects are used for each treatment.

Factorial ANOVA is used when the experimenter wants to study the effects of two or more treatment variables.

The tools covered in this chapter are standard in many industries including basic project management. The value of this chapter to the project manager is that the tools can be used to make projects, in general, leaner and more agile in nature.

It is important to remember that although there are a number of ways to analyze data, the seven tools of quality are often all a PM needs. This chapter serves as a good refresher of analytical tools as well as an easier way to explain the dynamics and purpose of these tools to the project team.

Chapter 13

Making the DMAIC Model *Leaner* and More Agile: Improve

Define, Measure, Analyze, *IMPROVE*, Control

In the Define, Measure, and Analyze phases of the DMAIC model, there is a lot of creativity involved, and many different choices and directions may be taken. As long as the major objectives are met, the project manager is working within the DMAIC framework. There are limited rules and mostly suggestions and ideas about how to mistake-proof the project.

The tools in this chapter are important to the project manager because many of them act as a double-check and are in one way, or another, related to the project plan regardless if the project manager is planning to use the formal DMAIC model.

Please note that once the project manager (PM) who has adopted the DMAIC methodology enters the Improve phase there is a specific step-by-step rule book that serves as an unyielding roadmap:

1. List the three to five solutions.
2. Be prepared to provide all crucial documentation.
3. Achieve agreement on which solution will be tried.
4. Perform a pilot.
5. Design a project plan.
6. Roll out the solution.

The PM does not leave the Improve phase until an improvement is made. Once implemented, the PM compares the before picture (Measure) to the after picture (Improve) to verify the improvement.

Key tools and activities include the following:

Brainstorming
Decision matrix
Pilot
Project plan
Failure mode effects analysis (FMEA)

The Improve phase includes a number of steps. The first step is to list the solutions discovered in the Analyze phase along with the research and logic (Figures 13.1 and 13.2).

The next step is to gain agreement on which solution to try, followed by a pilot or test (Figure 13.3).

The DMAIC suggests activities that can be done to accomplish the major objectives in each phase. The DMAIC does not promote implementing any solution without trying it out first (piloting) to eliminate mistakes. Whereas the pilot may take a number of creative approaches, it is essential that it is done (Figures 13.4, 13.5, and 13.6).

Solutions for improving the district marketing process for local offices
1. Have local offices continue to rely on staff for their local marketing activities, working in pairs. Process would require better communication, training, a way to document results and hold people accountable. Office managers are to research and assign staff for activities.
2. Hire a dedicated marketing ambassador—one per office who would work part time on busy weekends and partner with an associate within the office.
3. Hire a team of full-time designated marketing ambassadors at a district level who would travel to different offices during the week. Marketing ambassador would be assigned 3–4 office locations and be supported by the office staff. Ambassadors would also update staff on best practices, special offers, updates and track their progress.
Benefits of #3: Option to delegate marketing responsibilities: Each ambassador would be assigned a different role. For example, one for research and planning, one for planning and execution, one for training, one person for communication—reporting, building event website such as Google calendar or docs.
Another example cross marketing roles by type of field marketing: Assign one ambassador to all the chamber of commerce activities, one education to cover all the school organizations for example head starts across the area, one person festivals, one person partnerships example mobiles, group email with event alerts.

Figure 13.1   Example of a Solutions List.

The Six Thinking Hats

Solution option 1: Hire a professional marketing team to grow the business.

White	Data	Data sufficiently shows correlation between growth and use of new process during the pilot. The existing measurement tools from the current process are accurate, dependable, and easily accessed. Data shows that new process would cost less than current process and provide better quality.
Red	Intuition	Intuition shows that process would be simpler to execute than the current process, making the new process more Lean.
Black	From a negative point of view	Very high cost for marketing, especially if staff is not busy in the office. Large amount of responsibility for a small marketing team increases the chance of burnout and turnover. Risk of personality conflicts because the office manager will depend on both marketing manager and marketing team for ROI, increased revenue and new client growth.
Yellow	From a positive point of view	A professional motivated marketing team will bring in new clients and increase revenue.
Green	Creativity	Marketing is fun and exciting, and offers creative outlets such as collateral artwork, newsletters, event activities, contests.
Blue	Process control	Do we have the capability to keep the process in control and will the cost of a marketing team provide a high ROI.

Figure 13.2   Examples of Using the Thinking Hats.

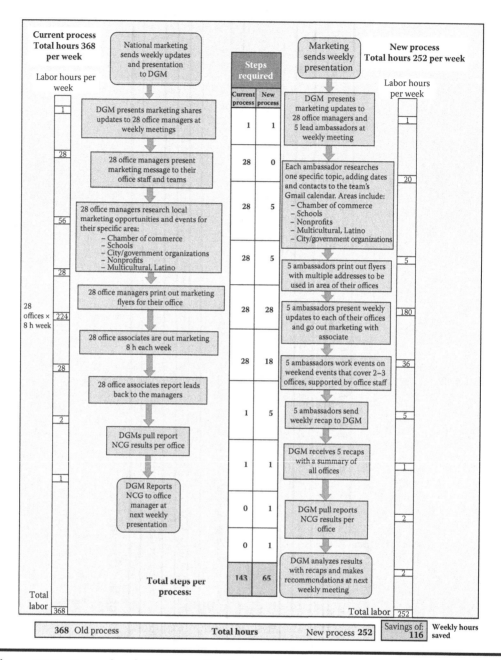

Figure 13.3   Example of Capturing the Current Process.

The final steps in the Improve phase include rolling out the project plan and determining the benefit of the project.

The target process of the Improve phase is designing creative solutions to fix and prevent problems in the future. This phase requires developing and deploying an implementation plan.

Trial periods results					
Trial period	**Trial period 1** **1/18 to 2/1**	**Trial period 2** **2/2 to 2/20**	**Trial period 3** **2/21 to 2/25**	**Trial period 4** **2/28 to 3/15**	**Trial period 5** **3/16 to 4/12**
District average YTD over prior year	−22.40	−16.24	−16.91	−10.5	−5.16
Sales during trial period	−5.38	−10.41	−16.79	−2.39	4.98
Increase in sales over prior year	17.02	5.83	0.12	8.11	13.65
Number of retail 1# offices with positive growth	2	8	7	8	18

Figure 13.4   Example of an Information Table.

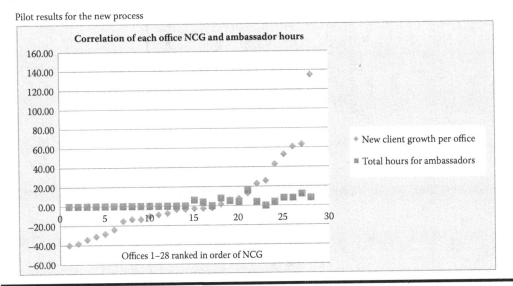

Figure 13.5   Example of a Correlation Chart.

Activities will involve developing potential solutions, defining operating tolerances, assessing the possibility of failures, and designing a deployment plan upon completion of a successful pilot.

The solution is rolled out in the Improve phase, so this is the most important phase to engage in mistake-proofing. FMEA (Figure 13.7) is one popular type of mistake-proofing tool.

Another popular mistake-proofing tool, poka-yoke, is used in processing, setup, missing part(s), operations, and measuring errors. The steps involved include the following:

Identifying the operation or process problem based on a Pareto chart
Analyzing the Five Whys and understanding the ways a process can fail
Deciding on an approach
Thinking about what might trigger this result
Trying out the solution
Training everyone

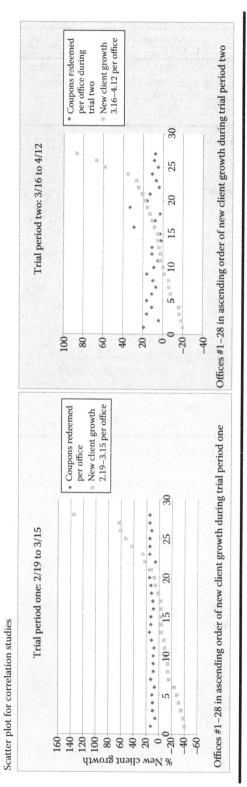

Figure 13.6   Example of Scatter Diagram Used in a Correlation Study.

Control process/product
Failure modes and effects analysis (FMEA)

| | Process or product name: | Marketing for new pizza restaurant | | Prepared by: Dawna Dougherty | | Page ___ of ___ |
| | Responsible: | Office manager | | FMEA date: 4/14/16 | | |

Process step	Key process input	Potential failure mode	Potential failure effects	SEV	Potential causes	OCC	Current controls	DET	Actions recommended	Resp.	SEV	OCC	DET	RPN
What is the process step?	What is the key process input?	In what ways does the key input go wrong?	What is the impact on the key output variables (customer requirements) or internal requirements?	How severe is the effect to the customer?	What causes the key input to go wrong?	How often does cause or FM occur?	What are the existing controls and procedures (inspection and test) that prevent either the cause or the failure mode? Should include an SOP number?	How well can you detect cause or FM?	What are the actions for reducing the occurrence of the cause, or improving detection? Should have actions only on high RPN's or easy fixes	Who is responsible for the recommended action?	Severe			
Local marketing for new pizzeria	Getting clients to come into the pizzeria to order pizza	Customers don't come in to order pizzas	Revenue decreases, new client growth goes negative. After a few weeks the pizzeria would go out of business	Severe	No strategic planning, lack of experience and training, very short staffed	80%	Manager provides a marketing plan and assigns his staff to execute	Not well	Find another way to get the message out to local customers	Office manager and staff				0
District managers to hire a professional marketing team to bring in new customers	Marketing team executes all local marketing #1 activities, sponsorships, and advertising	Marketing team does not provide any marketing activities	Customers are not reached and no pizza is sold	Severe	Team is not effective due to lack of planning, communication, #1 motivation, or accountability	20%	Professional marketing team has experience and is accountable. Weekly plan is already determined, daily report to be filled out, lead and appointment goals set per event	Well	Daily marketing report required, weekly calls with team members, rewards and contests for lead generation	District manager and marketing manager				0
Pizzeria staff will support and assist on local marketing events, as a subject matter expert to answer questions about the pizza	Manager to schedule one pizza employee to partner and shadow marketing team on each event	Manager does not schedule an associate to support marketing	Marketing team works alone. Activity is much less effective	Moderate	Communication failure, scheduling errors, staff not available or willing, poor leadership, personality conflicts	30%	Schedule confirmation between manager and staff, daily recap sheet have name and hours of staff to be entered	Very well	Send out a reminder email and phone call, rewards to staff who supports marketing	Office managers				
Pizza competitors' come out with new ad campaigns and huge discounts	Pricing—try to match competitor's offers	Cannot match competitor ads or discounts	Customer goes to competitor	Moderate	Company does not respond quick enough to match the discount, company will not match the discount, discount is not well understood by the customer	50%	Customer rewards program, social media, word of mouth including customers. Controls include marketing team's ability to act quickly and mobilize effectively	Well	Better communication by sending the updates directly to all employees including emails and post updates on the wall	National marketing to provide message. Local managers to offer discounts to match competitors discount				0

Figure 13.7 Example of a Failure Modes Effect Analysis (FMEA) form.

All mistake-proofing tools are concerned with anticipating what could go wrong and fixing it beforehand. The Pugh matrix, which is a decision matrix, may also be used. This involves compiling a chart: On the left-hand side, list the problems, and on the right-hand side, decide on a rating system. Having a matrix handy that shows criteria selection and solution rankings is a good quick way to answer questions. In fact, distributing this type of document in a meeting will keep questions relevant in the discussion.

Factors to consider in the Improve phase include the following:

Change management
Solution selection techniques
Criteria selection and solution ranking
Pilot planning
Pilot implementation schemes
Time management
Giving feedback
Communication skills
Managing stress

Change management in the Improve phase applies to managing the process and procedure that has been put in place to suggest a change, monitor the change, and evaluate the change.

The solution must always be subject to a pilot. How the pilot is implemented is of great importance. Were the right people included? How were the results measured? What was the true cost of the pilot?

Basic time management plays a key role in this phase. An implementation plan, the ultimate time management plan, is an expected deliverable. For those who have problems managing their time, a quick primer in time management techniques is recommended (see Figure 13.8).

Giving feedback, general communication skills, and managing stress are all part of the puzzle. Lean Six Sigma follows the belief that these issues are crucial to the successful implementation of any project.

On the more technical side of the Improve phase, the objectives are to do the following:

Perform a design of experiments (DOE)
Define operating tolerances of potential system
Assess failure mode of potential solutions

Phase	Days	Week 1	Week 2	Week 3	Week 4	Week 5	Week 6	Week 7
1. Content analysis	3							
2. Content collation	3							
3. Course analysis	1							
4. Course design	1							
5. Course writing	4							
6. Review and refine	2							

Figure 13.8   Example of a Simple Gantt Chart.

DOE includes the design of all information-gathering exercises in which variation is present whether under the full control of the experimenter or not. This is another area mostly addressed in manufacturing. DOE is a systematic approach to the investigation of a system or process. A series of structured tests are designed in which planned changes are made to the input variables of a process or system.

In the experiment, deliberate changes to one or more process variables (or factors) are made in order to observe the effect the changes have on one or more response variables.

The order of tasks to use this tool starts with identifying the input variables and the output response that is to be measured. For each input variable, a number of levels are defined that represent the ranges for which the effect of that variable is desired to be known. Several methods are available. The Taguchi method refers to techniques of quality engineering that consider both statistical process control (SPC) and new quality-related management techniques. This approach is favorable to Lean Six Sigma because it considers both management techniques and statistics in its approach.

Another area that has to be examined and refined in the Improve phase is defining the operating tolerance of the potential system. Is the system robust enough? Will it work well enough to meet any criteria previously set?

The assessment of failure modes is crucial. This can be done by using a simple FMEA as discussed earlier.

In the improvement phase, it is necessary to create and roll out the project plan. Some industries will refer to this as the execution plan or deployment. Some companies have methodologies that also include releasing the risk management plan or other documents in the same package of documents. The project plan is always done using work breakdown structure (WBS).

Before the rollout, it is essential that a pilot be performed. Simulation software is more popular in some industries than others, but it is a useful tool when a live pilot is not possible or too expensive. Sometimes a small focus group may be used to perform a pilot. Do not move forward until the idea is tested.

Even after the idea has been tested, mistake-proofing exercises need to continue. Once an outline of the project has been created, including all necessary steps to achieve the process improvement, each item on the WBS should be investigated. It is necessary to calculate the time and expense of each entry.

At this point, applying FMEA thinking might be appropriate. Checking on each key activity, making sure that a risk management plan is in place and that proper process controls are being recognized is essential. When the project plan is completed, before the baseline is created, it should be presented in draft form to all involved for review.

Tollgates in this phase primarily focus on the successful completion of the project.

Once an improvement has been recognized and documented, it is safe to move on to the final phase. As the project plan will still have closeout functions, it cannot be completed until the Control phase.

Project Plan

A good project plan effectively balances the components of time, cost, scope, quality, and expectations. Most experts agree these factors should be considered when designing a project plan. Taking all of these factors into account will help mistake-proof the project.

Definition/scope: The primary purpose of the project, including major functions, deliverables, and the purpose of the project relative to the organizational whole

Resources: The financial, technical, materials, and human resources needed

Time: Elapsed time and actual work time required to complete the project

Procedures: The various organizational requirements, policies, procedures, methodologies, and existing quality program

Change: New or different future conditions, requirements, events, or constraints discovered within or imposed on the project

Communications: Meetings, status reports, presentations, and details that
   will affect communication
Commitment: The degree of sponsor, user, and other stakeholder support
Risk: The potential constraints to project success

Designing a project plan requires listing all the steps in the process necessary for success. Each step is then assigned a resource, a timeline for completion, and a basic cost. Once the project plan has been reviewed, both a
time and cost baseline are created. This baseline is used from the beginning to the end of a project to determine if the project is within acceptable
parameters. Lean Six Sigma supports using WBS discussed earlier in this
text. Project plans may be developed easily in MS Excel although most project managers practitioners prefer using MS Project because MS Project offers
other applications useful in project management.

   Project plans need to be concerned with constraints (things that could
get in the way of project completion) as well as assumptions (things that
are assumed will be in place) in order to mistake-proof. Resource allocation
is always a major concern. A process improvement project plan follows all
the same rules as a typical project plan as supported by material produced
by the Project Management Institute. A good project plan should include
the overall expectations, definition, schedule, and risks of the project to
the organization as well as the blueprint (list of activities). The project plan
shows not only the project activities, but how these activities will be controlled throughout the project.

   This is a quick way to develop a draft of a project plan:

   1. Create a task list and WBS (Figure 13.9).
   2. Indent or out-dent tasks to finalize the WBS.
   3. Enter task durations or work estimates.
   4. Create dependencies between tasks.
   5. Assign resources.
   6. Pass the draft around for feedback.

Risk management should be one of the considerations in the Analyze phase
when compiling a list of possible solutions. It is particularly important in the
project plan phase. Risk management not only manages threats, but often
alerts the project manager to additional process improvement opportunities.

   The Improve phase, for an instructor, is often the easiest phase to teach.
This is because it is the strictest phase of the DMAIC model and really does

New process plan:

Scope: The purpose of this project is to increase new client growth over the prior year.

Level 1	Level 2	Level 3	Length of time hours	Total cost for the project
1 Milestone one: gain formal acceptance and approval	1.1 Cost analysis of new process	1.1.1 Calculate pay scale and salary for new marketing team with five ambassadors for 14-week period	2800	$32,000.00
		1.1.2 Estimate of weekly mileage between retail offices, to be grouped in close proximity	4	$3000.00
		1.1.3 Technical equipment including one iPad and company cell phone per team staff	1	$2800.00
		1.1.4 Training, uniforms, printing, marketing toolkits, and marketing inventory for events.	10	$3280.00
		New process estimated total	2815	$41,080.00
	1.2 Determine cost of current process	1.2.1 Marketing Hours: 10 hr. per week × 28 offices × 12 weeks	3360	$39,200.00
		1.2.2 Technical equipment with district includes 2 iPads and company cell phone for one mc, ave one hr. per office to enter leads	28	$860.00
		1.2.3 Office materials, printing, marketing premiums, collateral, event fees, chamber fees per office	144	$4700.00
		Current process estimated total	3532	$44,760.00
	1.3 Determine the savings or increased cost	1.3.1 Compare total costs: Even with addition of 5 full time employees each with an iPad and cell phone, the **savings of $3680.** Time and money saved by not supplying every office with its own set of collateral and marketing material reduced the cost.	717	**Savings of $3680**
	1.4 Submit new budget for approval	1.4.1 Send total cost and saving over current plan to regional and district manager	1	N/A
		1.4.2 Make adjustments as needed or move forward with new process	1	N/A
2 Milestone two: creating the ambassador team	2.1 Hire the ambassador team	2.1.1 Create job description	2	
		2.1.2 Research recruiting options: colleges for interns, online websites, referral from employees	6	
		2.1.3 Post job listings	4	
		2.1.4 Respond to applicants	4	
		2.1.5 Schedule interviews	8	
		2.1.6 Determine a team matrix with applicant talents (strengths: research, event experience, physical, training and sales experience)	2	CRITICAL
		2.1.7 Coordinate with HR to send out offer letters	1	
		2.1.8 Confirm new hire completion with HR	1	
		Total hours 30	30	
	2.2 Prepare for team training	2.2.1 Compile distribution list for team and upper management. Include a hierarchy chart to show positions and responsibilities	1	
		2.2.2 Secure location date and time for training	2	
		2.2.3 Notify team with meeting request for mandatory training date and location	1	CRITICAL
		2.2.4 Prepare training materials (PowerPoint for facilitator, print outs for team assemble binders, interactive games and prizes)	8	
		2.2.5 Prepare a marketing toolkit with marketing binder of area demographics, forms, leads, iPad, laptop and iPhone	12	CRITICAL
		2.2.6 Set up group Gmail account with alerts for local events, health fairs, nonprofit announcements using key words and shared calendar (to be used by the team)	2	CRITICAL
		2.2.7 Determine which offices will be assigned to each ambassador including one low performing	2	
		2.2.8 Identify which offices in each group which have largest bottlenecks and any offices that need to focus on Latino growth	1	
		2.2.9 Collect shirt sizes when confirming attendance	2	
		2.2.10 Order uniforms/shirts to be distributed at training (price calculated in group one)	1	
		2.2.11 Send out reminder email two days prior to team, to facility (for overhead, seating and payment options)	1	
		2.2.12 Arraign for lunch delivery from local restaurant, purchase snacks and drinks	1	
		Total hours 34	34	

Figure 13.9 Example of a New Process Plan Showing Milestones.

Milestone	Task	Description	Hours	Critical
2.3 Team kick off and cross functional training	2.3.1	Double check list and pack all materials	1	
	2.3.2	Arrive two hours early and check overhead, mic, seating and room temp, food arrangements, and sign in sheets	2	CRITICAL
	2.3.3	Assign each ambassador to one of the geographical area groups of five offices to cover	1	CRITICAL
	2.3.4	Assign each BA to one of the following marketing channels and events to the team's Gmail calendar for the entire district. Note each BA will later be responsible to add contacts and events to the team's Gmail calendar (to reduce rework).	1	
		BA 1—Chamber of commerce costs, contacts, fees, meeting dates, events	4	
		BA 2—All school calendars, contacts, including booster clubs and organizations	0	
		BA 3—City sponsored calendars, events and contacts	0	
		BA 4—Nonprofits organizations and Health Fairs	0	
		BA 5—Multicultural, Latino organizations and events	0	
		Total hours training meeting 9	9	CRITICAL
2.4 Establish the communication plan	2.4.1	Ensure communication plan—Confirm emails and contact information	2	
	2.4.2	Send out training recap to BAs and ask for feedback	2	
	2.4.3	Send training updates to management, office manager, marketing team and stakeholders including mini newsletter featuring new marketing team, photos and recap of successful training	4	
		Total hours 8	8	
3 Milestone three: Preparing the marketing activations plan 3.1 Preparing the 12-week process plan	3.1.1	Prepare preliminary schedules for weekly phone meetings, weekly recap and reports procedure	6	
	3.1.2	Each team member conducts 2 day research, adding results to districts Gmail calendar	80	CRITICAL
	3.1.3	Contact event organizers, pay and confirm events	10	CRITICAL
	3.1.4	Compile a rough draft of the 12 weeks activation from 1/15/15 to 4/15/15	4	
	3.1.5	Submit the 12 rough draft of the 12-week plan to regional and district manager	1	CRITICAL
3.2 Weekly ambassador schedule	3.2.1	Monday—Submit recaps, conduct research and reports, team conference call, email offices with marketing schedule 4 × 5	6	CRITICAL
	3.2.2	Tuesday—Wednesday days off unless special events are occurring	0	
	3.2.3	Thursday am—Team meeting. Pack inventory	4	
	3.2.4	Thursday 1pm–5—Activation by visit to office one, train the office team with news, event supports, results × 5	4	
	3.2.5	Friday–Visit 3 offices and train the office teams with news, confirm event support, results (2 hours per office visit) × 5	8	
	3.2.6	Saturday–Sunday—Activate at local events, visiting one office × 5	10	
3.3 Material inventory check	3.3.1	Check current inventory and order any missing items	4	
	3.3.2	Event set up materials check—full tent, weights, and complete inventory check sheet	2	
	3.3.3	Move larger items to a centrally located office for storage, with inventory check out procedure posted on the wall	2	CRITICAL
4 Milestone four: Execution of 12-week marketing plan 4.3 Weeks 1–3 execution	4.3.1	Focus on lowest performing office as determined by Pareto chart, spending 60% of time around that office	4	CRITICAL
	4.3.2	Conduct visits to other 4 offices, spending 2 hours per office	32	
	4.3.3	Record observations and make recommendations to marketing team and office managers (best practices, staffing)	4	CRITICAL
4.4 Mid season weeks 4–8 execution	4.4.1	Focus on middle performing offices, adding spring events to calendar and strong marketing candidates for event support	4	
	4.4.2	Conduct visits to other 4 offices, spending 2 hours per office	32	
	4.4.3	Record observations and make recommendations to marketing team and office managers (best practices, staffing)	4	
4.5 Late season weeks 9–12 execution	4.5.1	Refocus activities to the offices with the largest defects, checking for new opportunities and marketing assistance from office	4	CRITICAL
	4.5.2	Conduct visits to other 4 offices, spending 2 hours per office	32	
	4.5.3	Shift measurement to counting to a "new client countdown" determined for each office	4	
5 Milestone five: End of project wrap up 5.1 Final results and recommendations	5.1.1	Using the weekly tracker, sort the offices with highest growth and record the office marketing activities for those week	1	CRITICAL
	5.1.2	Record all marketing activity, document lessons learned. Make recommendations for next season	2	CRITICAL
	5.1.3	Update files of contacts and results into 12 week summary spreadsheet	1	
5.2 Inventory list	5.2.1	Update inventory list, returning any unused items to district storage, noting any missing or broken items needing to be replaced	4	
	5.2.2	Turn in all marketing electronics, including cell phones and iPads to district managers	2	
5.3 Budget wrap up	5.3.1	Confirm that all expenses and mileage have been submitted for payments	2	
	5.3.2	Analyze end of season budget and make recommendations, considering variation, money saved or wasted	4	CRITICAL
5.4 Celebration	5.4.1	End of season celebration party	2	
	5.4.2	Rewards and recognition for top performing teams and completion bonus	2	

Figure 13.9 (CONTINUED)   Example of a New Process Plan Showing Milestones.

not offer as much flexibility. Solutions are chosen, a project plan is developed, and the solution is rolled out. Each of the steps requires mistake-proofing. How much to mistake-proof the project plan depends entirely on the complexity of the project itself. For the project manager, regardless if they have embraced the DMAIC model, the value is learning tools that can enhance the project plan and create buy-in for the project.

Chapter 14

Making the DMAIC Model *Leaner* and More Agile: Control

Define, Measure, Analyze, Improve, *CONTROL*

In the traditional Define-Measure-Analyze-Improve-Control (DMAIC) model, an improvement has to be shown and documented in Improve first. This does not mean the project has been completed. There are a number of closeout activities that are performed in the Control phase. The Control phase is also responsible for showing a plan to sustain the process improvement.

There are two important things in this phase that project managers can embrace regardless if they are following the traditional DMAIC model. First is the documentation aspect of the project. Second, whereas sustaining the process improvement or project results is inherent to the DMAIC model, a project manager can use this phase to identify additional projects to complete.

The first activity in the Control phase, following the DMAIC roadmap, is to articulate the improvement in dollars. This is easy if the tool used to show the current process state in the Measure phase was financial. If the tool used to show the current state is not in dollars, a financial impact statement must be compiled. For example, in Measure, if the sigma level was 3.0 and the process improvement yielded a 4.0 what does that mean in dollars? If a scorecard was used in the Measure phase and the score was represented as a letter grade of "C" but after the process improvement the process is rated as an "A" what does that mean financially?

The second activity in the Control phase is to develop a sustainability plan. How can the company keep the process improvement in place? What are the red flags that should be watched? This may also include a transition plan.

Finally, all the activities usually associated with project closure should be considered:

Key tools:

Return on investment (ROI) formula
Sigma calculations
Control charts
Transition plan template

Tasks in this phase may also include the following:

Developing a transfer plan
Handing off the responsibility to the process owner
Verifying things such as benefits, cost savings, and potential for growth
Closing out the project
Finalizing the documentation
Celebrating!

The Control phase includes normal activities done whenever a project is closed, such as recording the best practices or notifying team members and the company that the project has been completed. Ultimately, the purpose of the Control phase has three basic characteristics.

The first is to verify and communicate the process improvement. This is usually demonstrated by showing the ROI or increase in sigma level. Keep in mind that the calculation for ROI and, therefore, the definition, can be modified to suit the situation.

ROI Calculations

ROI calculations can be easily manipulated to suit the user's purposes, and the result can be expressed in many different ways. When using this metric, it is imperative to understand what inputs are being used. The traditional formula is simple: ROI = net profit after taxes ÷ total assets. Another popular way to calculate ROI = the benefit (return) of an investment is divided by the cost of the investment; the result is expressed as a percentage or a ratio.

ROI Formula

Showing and documenting the process improvement is imperative. This needs to be demonstrated in a way that the new owner understands. Using sigma levels to show how the sigma has increased is also suggested. Control charts, discussed in the Analyze phase, are a quick way to help the new process owner watch the process stability.

Sustainability

The Control phase is used to communicate the plan for keeping the improved process in control and stable. In some companies, a control plan methodology or a control form may be in place. When creating the control plan from scratch, remember that the objective is simply the steps needed to keep the process improvement in place. The guide should be written in basic terms including the dates and the times when certain activities should occur. A control plan has two major components:

How processes are standardized
How procedures are documented

This may include a transition plan for the new owner. Even if the project manager plans to continue monitoring the process, this documentation is required. A transition plan gives the new process owner all the information needed to move forward. The transition plan is the document that would explain how to contact resources and how to use any tools presented in the control plan.

A strong transition plan includes the following:

An introduction
The scope
Transition activities
Roles and responsibilities
References and *attachments*

The transition plan provides the framework for identifying, planning, and carrying out activities. The purpose of transition planning is to ensure a seamless and continuous service when changing hands to new providers.

Example of Transition Plan

The deliverables in this phase focus on documentation. Processes are standardized. Procedures must be consistent. Transfer of ownership is established, and project closure is completed.

Transfer of ownership examines a number of areas. The purpose of the transfer of ownership document is to establish day-to-day responsibilities. It contains checks and balances to make sure the process continues to improve. A good plan would also include components such as knowledge and learning. It may even include job descriptions, staffing information, and where to locate future benchmarking data.

Success in this phase depends upon how well the previous four phases were implemented. A strong emphasis is placed on change management. The team develops a project hand-off process and training materials to guarantee long-term performance.

There are many factors that could affect the adjusted inputs and output. The process needs to stay in control, which is the most critical factor of this phase.

Showing and documenting standardization is an important part of the Control phase. Standardization enables high-quality production of goods and services on a reliable, predictable, and sustainable basis. Standardization is making sure that all elements of a process are performed consistently. Standardization allows the reduction of variation and makes the process output more predictable. It provides a way to trace problems and provides a foundation for training. It also gives direction in the case of unusual conditions. Standardization can even be the main objective, especially if the project was designed to meet ISO requirements.

Questions in the tollgate for control might be the following:

What process controls are being implemented?
Who is the process owner?
How often will the transition plan be revisited?
What is the expected improvement in the terms of cost reduction?

Finally, the Control phase is about developing and capturing best practices. The term is used frequently in the fields of health care, government administration, education systems, and project management.

An area that is addressed in the Control phase is a response plan. What are the critical parameters to watch? Is there a closed-looped system, meaning that nothing can fall through the cracks? Does a troubleshooting guide or frequently asked questions document need to be prepared?

In the Control phase, the document retention practices of the company should be considered. A document retention program involves the systematic review, retention, and destruction of documents received or created in the course of business. How the document retention program is implemented involves the balancing of potentially competing interests, such as legal obligations, efficiency considerations, and prelitigation concerns.

Some companies have an automated process. These systems partner well with the Lean Six Sigma theory because Lean Six Sigma is designed to eliminate waste and speed up processes. If these principles are applied to document retention it will yield immediate results. Electronic record management is a key component of success.

There are a number of compliance issues that govern how long documents should be retained for particular industries. Many companies face problems with document retention. This is an excellent opportunity for process improvement. How a company manages their documents can determine the future success of other process improvement programs.

The sheer volume of business information has increased over the past decade. Even email messages are now considered to be corporate records. Companies that are able to implement a successful document retention strategy realize significant savings.

There are several opportunities for process improvement:

Is there a reliable and valid indexing system in place?
Can documents (including email messages) be produced upon request?
Is there a data repository as well as a backup system?

5S Plans

As mentioned early in the book, 5S is a physical organizational system. Although 5S may be implemented before the DMAIC process begins, it may also be used to sustain the overall improvement or be used as a suggestion at the end of a project.

5S is a popular Lean Six Sigma tool that is designed to instill a sense of responsibility in employees and promote a disciplined approached. The original Japanese terms seri, seiton, seiso, seiketu, and shitsuke, used to describe the 5S model, are frequently replaced by a variety of English words. The attempt to develop English equivalents starting with first letter "S" has sometimes caused translation confusion for those trying to implement the model.

For example, almost all English translations of the 5S model will use the word sort as the first S, rather than the Japanese word seri. Seri is translated as "the identification of the best physical organization of the workplace." Seri (or sort) is often accomplished by discarding all unnecessary items. In English, however, the word sort is often used to mean "to place in different piles." The activity of placing items in different groups happens in the second S, known as seiso. Seiso is technically intended to arrange things in various piles or bins. In some models seiso, once again in keeping with the S theme, is referred to as systemic arrangements. In other models, it is called set in order. Shine, the third S, is referred to as spic and span in other models, which is a more accurate translation of the Japanese word seiso. In this article, the terms sort, set in order, shine, standardize, and sustain will be used to represent the 5S model.

The first opportunity to maximize success, when implementing a basic 5S program, is for the facilitator to clearly explain the definitions and use words that the employees identify with.

Some companies have decided to use the 5C model (clear out, configure, clean and check, conformity, customize and practice), which is very similar to 5S but has an easier vocabulary for English speakers to digest.

The next step in maximizing a basic 5S program is to study the company's infrastructure and decide how 5S can best fit in the existing improvement structure. This should be followed with constant, but brief, communications to the workforce explaining the 5S initiative. Several formats should be considered, such as email, electronic bulletin boards, and articles in the company newsletter.

The leadership team should be trained in the overall concept, and employees directly involved should be trained on each area of 5S. In the first step, sort, one of the main objectives is to discard unnecessary items. Employees should understand the criteria for making this decision. In the second step, one of the main objectives is called set in order, which means to place things in the right places. Will a color-coding system be used, or will a system be used in which the items most frequently accessed will be placed in the most convenient area? In the third step, shine, piles are revisited, reexamined, and often cleaned or refurbished. Once again, what are the requirements? For example, art items may have a certain way they should be handled. Chemicals may have certain safety criteria.

Educating employees in the standardization and sustain phase may be facilitated by a series of workshops but may also be satisfied with solid, easy-to-understand documentation. For example, in the fourth step of standardization, employees could be introduced to a diagram showing visual controls and be invited to discuss areas of risk. For the final step, sustain, employee training may consist of frequent updates on the success of the system through the company newsletter or targeted emails.

Clearly, one way to maximize the success of a 5S program is to ensure each employee has the appropriate amount of education. This would also include those facilitating the project. Facilitators and leaders of the 5S effort should have a strong understanding of project management and deployment plans.

The best way to gain buy-in to a 5S program is to start with a pilot that actually shows results. Select a small area or a neglected area that can show benefits within one week of implementation. All companies have a supply room or filing systems that could use a quick facelift. Cleaning up the supply closet is a simple way to visually show the benefits of the 5S program.

Before embarking on an enterprise-wide implementation, develop a full rollout plan and discuss with all parties involved. Once the rollout begins, be sure to collect best practices along the way for future projects.

Closeout Activities

Most of the closeout activities necessary in a process improvement project are the same as any project. These include the following:

Informing all parties (employees, vendors, etc.) that the project has been completed
Recording best practices
Updating documentation

However, every project is different. Closeout activities may vary. Often knowing this information as early as the Define phase can help the project manager know how to include these activities in the project plan.

Summary of DMAIC

DMAIC is an acronym for five interconnected phases: Define, Measure, Analyze, Improve, and Control. It is a Six Sigma business philosophy that employs a client-centric, fact-based approach to reducing variation in order to dramatically improve quality by eliminating defects and, as a result, reduce cost.

The Define phase is where a team and its sponsors reach agreement on what the project is and what it should accomplish. The outcome includes the following:

A clear statement of the intended improvement (project charter)
A high-level map of the processes (SIPOC)
A list of what is important to the customer (CTQ)

The tools commonly used in this phase include the following:

Project charter
Stakeholder analysis
Suppliers, inputs, process, output, and customers (SIPOC) process map
Voice of the customer
Affinity diagram
Critical-to-quality (CTQ) tree

The Measure phase builds factual understanding of existing process conditions. The outcome includes the following:

A good understanding of where the process is today and where it needs to be in the future
A solid data collection plan
An idea of how data will be verified

The tools commonly used in this phase include the following:

Prioritization matrix
Process cycle efficiency
Time value analysis
Pareto charts
Control charts
Run charts
Failure mode and effects analysis (FMEA)

The Analyze phase develops theories of root causes, confirms the theories with data, and identifies the root cause(s) of the problem. The outcome of this phase includes the following:

Data and process analysis
Root cause analysis
Being able to quantify the gap opportunity

The tools commonly used in this phase include the following:

Five Whys analysis
Brainstorming
Cause-and-effect diagram
Affinity diagrams
Control charts
Flow diagram
Pareto charts
Scatterplots

The Improve phase demonstrates, with fact and data, that the solutions solve the problem.
  The tools commonly used in this phase include the following:

Brainstorming
Flowcharting
FMEA
Stakeholder analysis
5S method

The Control phase is designed to ensure that the problem does not reoccur and that the new processes can be further improved over time.
  The tools commonly used in this phase include the following:

Control charts (covered in the Measure phase)
Flow diagrams (covered in the Analyze phase)
Charts to compare before and after such as Pareto charts (covered in the
  Measure phase)
Standardization

The control process involves quality and statistical concepts that have existed for decades. However, the advent of quality control software makes the process simple enough for anyone to perform.

Variation is everywhere, and it degrades consistent, good performance. Valid measurements and data are required foundations for consistent, breakthrough improvement.

Having a standard improvement model, such as DMAIC, provides teams with a roadmap. The DMAIC is a structured, disciplined, rigorous approach to process improvement consisting of the five phases mentioned, and each phase is linked logically to the previous phase as well as to the next phase.

Other benefits of using the DMAIC model often include the following:

Better safety performance
Effective supply chain management
Better knowledge of competition and competitors
Use of standard operating procedures
Better decision making
Improved project management skills
Sustained improvements
Alignment with strategy vision and values
Increased margins
Greater market share
Fewer customer complaints

In closing, whereas the Control phase of the DMAIC model is primarily concerned with documenting the success and providing a model for sustainability a project manager can benefit from the tools covered in this phase. Everyone can use a quicker way to document project success and the sustainability exercises can be used to create additional projects.

Chapter 15

Ethics and Social Responsibility

What Project Managers Should Know about Ethics and Values in a Lean and Agile Environment

When Alice, the heroine of the famed Lewis Carroll novel, decides to engage in the game of croquet, she is simply coming along to play. Everything goes terribly wrong, and the game is never completed because of the inconsiderate actions of the other player. Whereas the Queen, as well as other players, act in an outrageous way, the moral of the story is certainly that in order to accomplish things, people must work together respectfully and efficiently.

Ethics is motivated by the ideas of right and wrong. It is the philosophical study of moral values and rules. Business ethics also involves a company's compliance with legal standards and adherence to internal rules and regulations.

Values are the beliefs of a person or social group in which they have an emotional investment (either for or against something). Values are the ideas we have about what is good and what is bad and how things should be.

Business principles may be referred to using any combination of statements, such as mission, vision, values, or code of conduct. Without the proper moral and ethical framework in place, these statements are meaningless.

In fact, making these statements public can be more damaging than helpful to an organization that is not prepared to "walk the talk."

What is the difference between values and ethics? According to Frank Navran, "Values are our fundamental beliefs or principles. They define what we think is right, good, fair, and just. Ethics are behaviors and tell people how to act in ways that meet the standard our values set for us."

Too often, the subject of ethics and values is reduced to rules. Often rules are put in place that managers are allowed to bypass or dismiss. It is almost too simplistic to say that if a corporation's words and deeds don't match their ethics or code of conduct policies a values-based culture does not have the chance to exist. In its simplest terms, it is "what you do, how you do it, and when and what you say," that determines the foundation of the organization.

In the last decade, the term corporate social responsibility (CSR) has become popular. Generally, CSR is defined as "achieving commercial success in ways that honor ethical values and respect people, communities, and the natural environment." CSR means addressing the legal, ethical, commercial, and other expectations society has for business and making decisions that fairly balance the claims of all key stakeholders.

There are a vast number of councils, coalitions, international organizations, and private enterprises dedicated to CSR. CSR includes issues related to business ethics, community investment, environment, governance, human rights, the marketplace, and the workplace.

Ethics

Warren Buffet was asked what the three key attributes of corporate leaders were. He said, "integrity, intelligence, and energy—without the first, the other two will kill you."

Ethics is the standard of conduct that guides decisions and actions, based on duties derived from core values. Ethics, integrity, and trust all start at the top. However, the initial discussion regarding these topics often originates in the human resources (HR) department.

Both the NYSE and the NASDAQ require a "code of business conduct and ethics" covering all employees, officers, and directors.

The Sarbanes–Oxley Act requires a public company to disclose whether it has an ethics code for senior financial managers, which includes the CEO. Having training in place that addresses these employee responsibilities would help minimize exposure.

Compliance with this act is very serious. Some larger companies claim that it costs an average of $16 million each year to comply with regulations. Sarbanes–Oxley imposes criminal penalties for corporate governing and accounting lapses.

There are key questions that should be considered before embarking on any discussion in this area, such as the following:

Do managers know and believe in the corporate mission?
Are managers committed to the mission?
How do managers demonstrate their commitment?
Usually, the HR role in the discussion of ethics is that of facilitator.

HR might also develop the message and/or be charged with deploying and delivering the message. Too often, this is done by simply providing documents that must be read and signed by the employee. Ethics programs can be as simple as signing a document or as thorough as discussions, follow-up, and evaluations via focus groups.

Developing a set of principles that a company can live by (that honestly guides decision making) is no simple task. It involves soul searching. Many companies are not prepared to engage in this type of activity.

Traditionally, ethics policies have been inserted into employee handbooks or presented as a separate document to new employees on their first day of employment. Sometimes, this is the extent of the training. However, in recent years, there has been a stronger emphasis placed on the subject of ethics. Project managers should take a proactive approach and reexamine these policies.

A new trend involves consultants and self-employed contractors writing their ethics policies and including these statements in the contract.

A good policy or procedure may enhance culture once it is deployed. A clearly stated and published policy of required and prohibited employee activity will create a stable working environment.

A confusing policy may have the opposite effect. Policies may also be abused, distorted, and neglected. Most of the exposure in ethics policies results from policies that have been in place for a number of years and that have never been reviewed or revised. Policies need to be updated periodically because state and federal laws continue to change.

The term workplace ethics has become popular. This catch-all term can cover anything from sexual harassment to civil rights infractions to privacy concerns. Workplace ethics are the standards by which a company dictates how employees should treat each other and the business. Managers may also hear the term business ethics, which refers to a company's attitude and conduct toward its employees, customers, community, and stockholders.

Workplace ethics topics may include discussion regarding race and gender. Involuntary stereotypes affect the way we deal with other people. Often attitudes, such as ridicule, put-downs, and accusations, normally discussed in classes on harassment are finding their way into discussions regarding workplace ethics.

Global and international ethics policies are designed to speak for the company. There are specific industries, such as health care, that must consider the impact of their statements when working with different cultures and religions.

As a project manager, the role may be limited to having the employee sign a statement that they have read the ethics or code-of-conduct policy. However, it is important to understand and to be able to explain the company's program.

Many public companies have taken the opportunity to update their code of ethics partially in response to the Sarbanes–Oxley Act of 2002—a law designed to improve corporate accountability. It is important as a project manager to stay abreast of and understand new developments.

Due to the many corporate scandals, recent graduates are inquiring about company ethics during the initial interviewing process. Increasingly, even seasoned applicants are concerned about corporate ethics policies.

A company's base philosophy may also be a consideration for many candidates. In these cases, recruiters for publicly held companies can demonstrate how compliant the company is with regulations. Privately held companies can stress stability in the market and tie it to their ethical practices.

Ideally, a company should talk about its ethics and company philosophy when presenting the history of the company to the applicant.

As long ago as Aristotle's *Nicomachean Ethics* and Plato's *Meno*, there have been two moral questions posed. One involves action and the other involves character:

How should I act?
What type of person should I be?

Can adult character be developed? This is a debate psychologists have never settled. Most psychologists will agree, however, that there are four major questions to be addressed when focusing on character development:

What is good character?

What causes or prevents it?

How can it be measured so that efforts at improvement can have corrective feedback?

How can it best be developed?

Psychologists generally agree that the factors influencing character development the most are the following:

Heredity

Early childhood experience

Modeling by important adults and older youth

Peer influence

The general physical and social environment

The communications media

What is taught in schools and other institutions

Specific situations and roles that elicit corresponding behavior

Individual character development education is the long-term process of helping individuals to be motivated to live by a set of ethical standards. These programs are often too complicated and time-consuming for businesses to adopt. Rehabilitation is certainly a type of character development education.

Corporate character development programs, such as peer reviews, lunch discussions with managers, and participating in case studies, do appear to be effective in refining behavior.

These programs also might include a formal code of ethics or code of conduct statement.

In addition to simply scaring applicants, corporate scandals in 2002 and 2003 provide a stark reminder of the catastrophic risks involved with business ethics failures. Critics of formal ethics training courses will say that ethics is not a program but rather a group of habits. There is a tendency to direct ethics and ethics training conversations toward the executive team; however, ethics impacts everyone.

Providing employees with incentives that are solely linked to performance or financial success often leads to unethical behavior. The fear of not meeting the bottom line can cause even the most honorable employee to stray. Tangible rewards must exist for those who approach business from an ethically aware and ethically inspired focus. It is important to create a culture in which ethical behavior is the only option.

Recognition in general is an important component of an organization's total rewards program. There is evidence validating that rewards programs help in reducing turnover, increasing productivity, and creating a positive work environment.

It is important to recognize that the scope of business ethics has expanded to encompass a company's actions with regard not only to how it treats its employees and obeys the law, but to the nature and quality of the relationships. Ethics policies must also consider the following:

Shareholders
Customers
Business partners
Suppliers
Community
Environment
Future generations

Online ways of doing business have created new ethical dilemmas. Some businesses use tools such as anti–money laundering auditing checklists or computerized complaint procedures for accounting or auditing issues. But the fact remains that instituting integrity and developing an ethical culture is the best risk management plan. In today's environment, business ethics and integrity are under a microscope. The stakes are high. Companies who stay committed to their values will prevail.

Values

There is considerable confusion surrounding the definition of values. Kurt Baier, noted philosopher, stated that sociologists employ a bewildering profusion of terms, ranging from what a person wants, desires, needs, enjoys, and prefers through what he thinks desirable, preferable, rewarding, and obligatory to what the community enjoins, sanctions, or enforces.

For an individual, values may be faith- or religion-based. But typically when the word value is applied to business, it can be defined as the company's sense of character or integrity.

Because managers in the same workgroup may define this "sense" differently, it is not always easy to identify or capture in writing.

In *Building Your Company's Vision* by James Collins and Jerry Porras, core values are described as the essential and enduring tenets of an organization—the very small set of guiding principles that have a profound impact on how everyone in the organization thinks and acts. In the authors' words, "core ideology provides the glue that holds an organization together through time."

Values are what set companies apart from one another. These values should be remembered because they are core to how the company thinks, but they should likewise be inspiring. For example, the Build-a-Bear company refers to its values as "bearisms." These bearisms are embraced by the employees. The mission statement:

> At Build-A-Bear Workshop®, our mission is to bring the Teddy Bear to life. An American icon, the Teddy Bear brings to mind warm thoughts about our childhood, about friendship, about trust and comfort, and also about love. Build-A-Bear Workshop embodies those thoughts in how we run our business every day.

Many companies list their values as something that sets them apart. Other companies print and post their values. Many business values center on customer focus and integrity. Whatever the corporate values are, they should be recorded and distributed to employees. When possible, discussion around company values and activities that enforce these values should be enacted.

The project manager will be looked at as the go-to person if there is no official HR representative.

Most companies make decisions according to a few core values. Some are written down, and others are implied. A written values statement communicates what's important. The statement makes it clear what managers believe. Values help people embrace positive change.

For managers assigned to the task of determining or validating values, these simple points should be kept in mind:

In what do managers believe?
What really matters to managers?
What values would managers like to pass on to present and future
  employees?
What values really govern behavior?

Leaders cannot be successful if they don't support or believe in the values they promote. Without respect for the company's values, a leader will fail even with the best training.

For managers to demonstrate a commitment to company values, their professional profile must remain high. They should be aware of their own personal preferences and understand how these preferences may impact their decision making.

Often the improvement cycle starts with the organization's vision, values, and purpose. We look at everything differently today that may result in a change in values and a reexamination of what is ethical.

Thirty years ago, topics such as globalization, graying of the workforce, and the impact of terrorism were not part of our physical makeup. Telecommuting, technology, unemployment and even the multitude of mergers and acquisitions have changed our core thinking.

There is a great deal of attention placed on generational differences. But workers of all ages may have more commonalities than differences. Most workers would agree that honesty and integrity are vital to successful leadership. Both groups place a high premium on workplace respect.

One generational difference that does cause tension between older and younger workers is the question of work/life balance issues. As a general rule, younger workers and women value work/life balance opportunities more than older male workers who place more of a premium on job satisfaction.

Not long ago, business ethics was dedicated to compliance-based, legally driven codes. Training outlined, in detail, what employees could or could not do. Most information was focused on conflict of interest, expense reports, or improper use of company assets.

In the new economy, companies are creating values-based, globally consistent programs. These programs help employees make sound decisions even when they are faced with new challenges.

When thinking about the topic of this chapter, ethics and values are important to remember. The first step is becoming ethically aware. Evaluate the current ethical climate.

Next, it is important to define values. Identifying, testing, and ranking these values cannot be a one-person job. It requires time, employee participation, and feedback.

Ethics has become a popular topic, not only in project management, but in other areas of business as well. Currently, most scholars claim that there are three schools/thoughts of ethics. However, other individuals claim that there are five different areas of ethics and some as many as eight.

They are classified as virtue ethics, consequentialist ethics, and deontological or duty-based ethics. Each approach provides a different way to understand ethics. For project managers wanting to study more about ethics, in general, the Leaner approach is to study the three-type model.

Chapter 16

The Lean and Agile Project Leader/Manager Model

Being Both a Leader and a Manager

Project Managers are already managers of projects and people. Only recently were they also considered leaders. A good project manager will have enough leadership ability to inspire the team to complete projects.

The Lean and Agile project manager (PM) will benefit from becoming familiar with the work of W. Edwards Deming. Deming was a pioneer in the quality movement and is well-known for a number of works and theories related to process improvement and project management concepts. One of his most famous writings is the system of profound knowledge (SoPK).

This theory states that the following things are key to good management:

Appreciation of the system
Knowledge of variation
Theory of knowledge
Understanding of psychology

These areas are fully developed in Deming's work, but a summary follows of the four elements.

Appreciation of a System

A business is a system. Action in one part of the system will have effects in the other parts. We often call these "unintended consequences." By learning about systems, we can better avoid these unintended consequences and optimize the whole system.

Knowledge of Variation

One goal of quality is to reduce variation. Managers who do not understand variation frequently increase variation by their actions.

Theory of Knowledge

There is no knowledge without theory. Understanding the difference between theory and experience prevents shallow change. Theory requires prediction, not just explanation. Although you can never prove that a theory is right, there must exist the possibility of proving it wrong by testing its predictions.

Understanding of Psychology

To understand the interaction between work systems and people, leaders must seek to answer questions, such as the following: How do people learn? How do people relate to change? What motivates people?

In the United States, we generally identify leadership and management as separate roles. In other countries, the lines may be blurred. The Lean and Agile PM must be equipped to both lead and manage the process improvement effort. Understanding how different cultures react to authority will assist the Lean and Agile PM in understanding what role is most beneficial to the success of the project.

In some countries, such as Argentina and Brazil, it is crucial that managers act like managers. Leaders and managers should not try to communicate as though they were staff members. Then again, in Australia and Canada, it is just the opposite. Managers are characterized as informal and friendly.

Basic work ethics vary as well. For example, in Austria, team members want clearly defined instructions with the privacy and the confidence to pursue their goals without interference. So simple project tracking might be seen as a nuisance.

In Belgium, there is usually a desire to compromise. In China and South Korea, the typical business is still grounded in a Confucian philosophy. For example, an older person's thoughts may be more respected than a younger person's opinion. These cultural differences can be a challenge for the Lean and Agile PM.

Assessing the leadership model of the company and how much authority will be given to the Lean and Agile PM is just as important as understanding how employees will react to the new leader or manager. Even when the Lean and Agile PM is expected to perform in a project management role only, some leadership skills will be necessary to move the team forward.

Successful process improvement projects require both stellar leadership and solid management skills to survive in today's economy. As Lean and Agile PMs deal with limited resources, maintaining goodwill and the confidence of the team is necessary for a positive project outcome.

Building relationships with the team is critical. Even so, communication styles vary from culture to culture. Generally, in Malaysia, the communication style is very polite and diplomatic. In Italy, good communication is usually loquacious.

So, the challenge for the Lean and Agile PM is multifaceted. How does the team respond to various leadership and management models? How should basic communication be handled within the team? These considerations impact everything that the Lean and Agile PM is required to accomplish when implementing a process improvement.

Even in the Define, Measure, Analyze, Improve, and Control (DMAIC) model, certain cultures will react differently to each phase. In many cases, the Lean and Agile PM must become accomplished at seamlessly moving from one phase to the other without a lot of fanfare. In other cultures, such as Denmark, it is essential that everyone feel his or her value is included. Keeping everyone informed is significant. So, in this case, it is important to share progress with the team via the use of tollgates. Letting everyone on the team know when certain deliverables and milestones have been achieved is critical.

Specifying realistic goals and using common sense when determining available resources for the team is the responsibility of the Lean and Agile

PM. When leading the process improvement effort, relating these goals to the team and using the right level of sensitivity must be considered.

As noted earlier, in some countries, age is respected. Giving vital tasks to a younger employee may be seen as disrespectful. In the United States, achievement is generally respected more. Time is money in the United States. In other cultures, the relationship factor must be considered to get the process improvement project executed.

Warren Bennis said the way a person can differentiate between managers and leaders is that leaders "do the right things" and managers "do things right." The biggest difference between managers and leaders is the way they motivate the people who work or follow them. This sets the tone for other aspects of what they do. In most process improvement efforts, the Lean and Agile PM must take on both roles. However, to do the right things and do things right require a great deal of cultural communication skill.

Some Lean and Agile PMs come by both leadership and/or management competencies naturally. It will greatly benefit the Lean and Agile PM to consider their strengths and weaknesses. The Lean and Agile PM who is aware of his or her strong leadership skills may need to step back when tasks are not completed and return to basic management techniques. A Lean and Agile PM who is getting deliverables met but notices that employee morale and inspiration are starting to slip may need to move into leadership mode. This attentiveness and flexibility is hard enough to master in one's own culture.

In many countries, the manager has a position of authority, vested in him or her by the company. Subordinates largely do as they are told. On the other hand, when Lean Six Sigma projects are in startup mode, the Lean and Agile PM needs followers more than subordinates. Followers are inspired to do what the leader requests regardless of whether or not the leader has direct control over his or her position.

Some studies suggest that managers are risk-averse whereas leaders appear as risk-seeking. Many cultures will not immediately understand that Lean Six Sigma is all about mistake-proofing and reducing risk. Managers focus on the process and immediate efficiency more than leaders do. Leaders think about how they invest their time to develop the strongest talent so that those people can grow and do more and more over time. Again, the Lean and Agile PM must learn to do both.

Motivation, recognition, and incentive programs can assist a Lean and Agile PM to accomplish this objective. Some cultures prefer a collective award or no award at all. In other cultures, the role of the team is such a core element of employee identification that seeking out an individual contributor for recognition would not be welcomed. Many employees might be reluctant to step forward or to be pushed into the spotlight.

Leaders are charged with building a climate for employee commitment. This is achieved by aligning employees to the mission, vision, and value of the company. Ensuring that everyone understands how his or her specific role impacts the company and that each employee is encouraged to develop competencies is also a key to leadership.

The ability to engage employees is a crucial element of productivity, creativity, and success. Understanding communication styles and bias is necessary. Being a leader—especially at the top levels of any company—is never easy. Leaders are expected to master, not simply learn, new skills necessary to run the business.

Sorting out the uncertainties and the politics of the business can be daunting. The length of time a leader is given to "get it" is much shorter in the new economy. Results are often expected immediately. The best leaders have the ability to make people feel appreciated. This inspires loyalty. Good leaders maintain control over the key decisions without micromanaging their staff. They know where their energy should be spent, and they avoid solving problems better solved at a different level.

Bill Swanson, former CEO of Raytheon, believes there are three qualities of leadership: confidence, dedication, and love. "If you watch true leaders," Swanson states, "they're willing to do unbelievable things for the success of their teams or organizations. They have a passion that people sense."

Strong leadership is necessary to accomplish this task. Leaders must be willing to build alliances with employees. Lean and Agile PMs have more exposure in this area. Because projects are designed to produce results in a short period of time, it is critical to gain trust and respect very quickly.

Leadership development is increasingly regarded as the platform needed to grow and improve the business. Leaders must be able to deal with the uncertainty of global markets, competition, and the economy. Leadership development also addresses how to achieve strategic goals, organize innovative projects, and change culture. Finally, leaders must be able to consider the entire business, using technical, human resources, interpersonal, legal, and empathy skills.

A critical skill for leaders is the ability to manage their own education. Surveys on leadership indicate that the best characteristics of a leader include the ability to inspire individuals and to explain the vision.

Leadership skills have become more dynamic. Leaders must make decisions quickly and act within shorter time frames. They must master puzzles and know how to ask the right questions. Successful leaders must be able to handle conditions of ambiguity, complexity, and risk.

Successful leaders are stellar communicators with excellent presentation abilities. They continually improve their interpersonal skills, display good judgment, and maintain confidentiality. Leaders make business decisions that focus on business results.

The Lean and Agile PM may find it frustrating to know that there is no consistent proven approach to leadership development. There are many different perspectives and methods.

Most development practices focus on selected aspects rather than the whole. Problems stem from the fact that leadership in organizations is the result of a complex system of interactions.

Interactions include the leader's own personality, characteristics of the follower, demands evolving over time, and a host of environmental factors. When employees are exposed to good leadership, it is easy to identify. Capturing that ability well enough to provide training is a different matter. Many believe that leadership is intuitive. There is evidence that some leaders are simply born with the ability. However, there is still strong evidence that many leadership skills may be learned.

There are conditional demands that require leaders to align and realign their behavior. Excellent leaders must motivate and energize people to change. It would be too simplistic to say that true leaders must have the desire to do this, but in reality, leadership ability is heavily rooted in personal philosophy.

The Lean and Agile PM should first assess existing competencies against those desired and base development and training opportunities on the gaps. This study may also include specific behaviors. The gaps should expose or validate the following:

Is there a gap in the level of leadership?
Is there a problem with methodology or process?
Is it a philosophy issue?
Is it a simple inability due to training or the ability to perform?

By providing diagnostic information about various dimensions of leadership, a systems framework can clarify major areas requiring development. Gap analysis practices are generally accepted and easy to implement, but it is imperative not to rely only on this information when trying to discern what components should be included in a leadership development program.

Measuring performance gaps can be threatening to employees. It is valuable that the ideal performance is established and defined. Performance measurements consider quality, costs, benefits, product, and productivity. Even with all the diverse approaches and theories to developing leadership programs, it may be simple observation skills that best help a Lean and Agile PM determine which leadership program is necessary for a particular project.

Being good at observing will help any Lean and Agile PM determine a better leadership development path while improving his or her own leadership ability. Enhancing observation skills will assist in other areas as well. Here are a few tips on how to observe:

Slow down and watch.
Pay close attention to the physical surroundings. Be aware of people's reactions, emotions, and motivations.
Ask questions that can be answered thoroughly.
Be yourself.
Observe with an optimistic curiosity.
Be ethical.

In *The Feiner Points of Leadership*, Michael Feiner addresses what he refers to as basic laws that will make people want to perform better for managers. A thought outlined in Feiner's book is called "the law of intimacy." The law refers to taking the time to learn about the people on the team. Feiner points out that it is not necessary to be best friends with employees. Feiner believes that it is significant to listen to an employee and understand what is most important in his or her life.

Based on Feiner's thoughts on strong leadership, a Lean and Agile PM may notice that a particular person could use help in connecting with his or her employees. This observation might lead to mentorship opportunities over a lunch discussion rather than a formal class on communication skills.

There are a number of theories about what will make a leader successful. The strategy to design a leadership development program depends on the organizational culture or philosophy.

Organizational culture defines corporate strategy, how the organization may react to crisis, and the management style. Leaders need to understand that commonly held beliefs about the group and history are worth taking time to understand. Leaders must be aware of any symbolism in the workplace and investigate the meaning.

Leaders must not only be aware of subcultures, but also the countercultures. Subcultures are those that question basic assumptions and confront the central culture.

Countercultures can be created by any number of issues and may include the following:

Mergers
Social issues
Employee discontent

Cultures share many things in common. Collectiveness is a set of commonly held beliefs. How a group collectively feels about a particular issue and how they react are key to understanding culture.

In *The Third Option*, by Saj-Nicole Join, leadership skills are addressed as habits:

The habit of mind
The habit of relationship
The habit of focus

Leaders today must master a higher level of thinking and focus on the essential issues. They must assemble modern leadership teams.

Many psychologists believe that we adopt habits during our childhood by observing our parents and other adults. If this training was inferior in any way, then finding a mentor who demonstrates good leadership skills is necessary. Leaders need to develop habits that include the following:

Reflection
Framework
Attunement
Conviction
Replenishment

Reflection is the capacity to examine one's own behavior. Framework refers to the ability to create an optimistic narrative. Attunement is the practice of setting aside old beliefs and assumptions. Conviction is the ability to trust, value, and act/speak from one's own experiences. Finally, replenishment is the act of restoring perspective and the renewal of resources.

There is always a temptation to define areas relative to leadership development. Most experts agree that the following areas are imperative:

Vision or strategy
Courage
Understanding
Respect and trust in the management team
Decision making
Ability to generate both personal and organizational energy

Leaders must lead by example and be able to develop and deploy winning strategies. Leaders have the ability to inspire employees to achieve greatness.

Many works exist on how Winston Churchill, Jack Welch, Colin Powell, and others handled their leadership roles. Taking time to read this literature will not be wasted. Each leader has a message, and there is always an opportunity to learn new methods. Many great leaders had in-depth experience working with other cultures.

Leaders have acted as heroes, actors, power brokers, and ambassadors. We look to heroes to diminish anxiety and to save the day, raising our own comfort level. Heroes may present themselves as playful or warrior-like. Some leaders think of leading as a performance art. These types of leaders may act as poets, teachers, storytellers, or showmen. The Lean and Agile PM can learn from all of these disciplines. The better-rounded the Lean and Agile PM is, the better he or she will adapt culturally.

For all leaders, it is a question of increasing personal credibility. Leaders must be competent in relating to and communicating with diverse groups of employees by participating in trust-based strong relationships. They must be skillful at communicating key organizational messages.

Leaders at all levels must deal with issues of power and/or insecurity. In order to motivate, reward, and develop employees, it is necessary to secure support from the next level.

Examples of unsupportive behavior include the following:

Shifting goals
Confidence betrayal
Negative politics
Micromanaging

Lack of integrity
Not setting clear expectations

The most positive and beneficial approach is to exercise influence by improving relationships with the boss. Try to separate feeling from the facts. Negative confrontation rarely helps the situation. The best tactic is to keep the lines of communication open and frequent. Use an assertive approach when expressing needs.

Strategically, leaders should assess the situation and invest in solid self-assessment. Build value through people. Consistent earnings can be achieved by building credibility and communicating frequently. Having quick access to data is also necessary. True leaders realize that there is more to leadership than improving the financial picture. Leaders must also do the following:

Deliver consistent and predictable earnings
Articulate a future growth vision
Align competencies to strategy and create capabilities

Consider key avenues to growth, such as the following:

Innovation
Geography
Customer base

In summary, the Lean and Agile PM needs to engage and invest in self-study activities centered on continually improving management and leadership skills. The Lean and Agile PM then needs to identify both strengths and weaknesses in his or her style and consider how each strength or weakness impacts the culture. Finally, the Lean and Agile PM needs to be aware of when he or she needs to switch from a leadership role to a management role and vice versa.

Project managers today are often responsible for more roles than they were in the past. They may have to assume duties associated with the Human Resources or Accounting Departments. In many cases they must not only manage a project but lead the project. They may be the liaison to the C-Suite, a presenter, or an internal marketer to get the job done. Reading and studying books on leadership or taking courses in leadership development serve the project manager well since they are usually already versed in basic management practices.

Chapter 17

Change Management Basics: Lean and Agile Project Managers

Beliefs and values evolve with a company's history. They are not easily abandoned. Change that does not address or respect these values and beliefs will most likely not be successful.

In order to function in a change management role, it is important to do the following:

Stay current on the organization's mission, policies, and plans.
Be prepared to communicate those mission, policies, and plans.
Act as a buffer between executive management and employees during
   stressful mergers, layoffs, or changes in direction.

Although change is the inevitable reality, framed correctly, it is possible to return to the past for inspiration. A mistake in basic change management is to discount the old way of doing things entirely. This isn't necessary and can be counterproductive.

Employees cannot be considered malleable material when it comes to change management initiatives. It is impossible to handle people like a sculptor molding clay into various forms. There is pushback and resistance even when the change is ultimately a positive one.

A surprising number of employee relations issues can be directly traced back to how well change was presented or handled within a group. There was a time when change was temporary and always followed by a longer period of stability. Now, change is continual and does not allow employees time to regroup and accept the change before an even newer change is imposed.

The way an employee approaches the thinking process can determine how well that employee will adapt to change. Thinking outside the box, thinking analytically, and thinking holistically are all indicators that the employee will be able to adjust quickly.

Individual skills and competencies can position individuals to accept change better than others as well, and these include the following:

Technical ability
Understanding project methodology
Ability to create solutions
Capability to form partnerships

Without natural competencies and skills, individuals exposed to change may benefit by learning about and implementing systems thinking. This thinking involves five easy steps:

1. Stating the problem
2. Telling the story
3. Identifying the key variables
4. Visualizing the problem
5. Creating loops

Stating the problem is the first step in almost every methodology. In areas dealing with change it is crucial. Having the employee put the problem in story form helps the employee identify more closely with the issue. Variables are components in the story that may change over time. Variables may include things such as a change in management. Visualization of the story in graphic form sometimes helps detect the change or behavior necessary. Finally taking the story and illustrating which factors influence other factors is called looping.

There are two types of loops:

Reinforcing
Balancing

Reinforcing loops are self-fulfilling prophecies, either positive or negative. Balancing loops keep things in equilibrium.

A collaborative approach to change almost guarantees high participation, strong commitment, and the creation of a reasonable standard that may be measured for results.

A mandate from executive management is often given to the human resources (HR) department. This may include delivering training and composing communications relative to change management. There are a number of things that can go wrong with this approach, but the worst culprit is hurried communication.

Lean and Agile project managers are now expected to take an active role in the change management process. This participation is expected even when a program has been developed by HR, but in the new economy, even the formation of such an effort may become a duty of the Lean and Agile PM.

This can be challenging for a Lean and Agile PM because the executive leadership team may very well expect a metamorphic change to occur overnight. In an effort to hurry things along, some Lean and Agile PMs have been tempted to use email when a face-to-face meeting would be more appropriate.

Once a change or a change management plan is formed, it is necessary to discuss the balance of communication. Even in companies that have a multitude of meetings, there is often not enough communication provided regarding change.

Many employees are only concerned with how their day-to-day lives will change. Giving too much information can be overwhelming. Other employees fear a loss of position or territory.

Groups need their territories, and that it is one of the ways they define themselves as groups. The concept of group definition is important to consider when addressing change. It is important to remember that people will want to protect their territory. This is natural and should be expected. When managers or leaders use words such as reorganization, and employees realize that this means a redistribution of territory, it is unreasonable to think that some problems won't surface immediately.

A popular meaning of the term managing change refers to making changes in a planned and systematic fashion. Rather than allowing change to occur naturally, and often randomly, change management assumes that it is possible to introduce planned change and steer its development.

It is important that Lean and Agile PMs learn to embrace and demonstrate good change agent skills. Internal changes may be triggered by events originating outside the organization—environmental change—which is out of the manager's control. Implementing new methods and systems in an ongoing organization takes patience.

It is helpful to think of managing change in the same light as basic problem solving because it is a matter of moving from one state to another just as problem solving moves from the problem state to the solved state.

As a Lean and Agile PM, it is important to understand the company's philosophy to make a positive impact on change. It is important that the Lean and Agile PM's objectives embrace and align with the philosophy of the company. Some costly mistakes can be made by a Lean and Agile PM who has not been involved in the change management process. Areas to watch include the following:

No systematic plan
Under-communicating the vision
Declaring victory too soon

One of the best-known philosophies was that every GE business had to be No. 1 or 2 in its market according to Jack Welch. Otherwise, it should be fixed, sold, or closed. Introducing change that does not mesh with the philosophy of the company will not be successful.

One of the best resources on change management is the Wharton Center for Leadership and Change. Works written by William Bridges, considered the pioneer of change management theory, are also valuable references to read or review before implementing a change management plan.

The mission of Wharton's Center for Leadership and Change Management is to do the following:

Stimulate basic research and practical application in the areas of leadership and change.
Foster an understanding of how to develop organizational leadership.
Support the leadership development agendas of the Wharton School and University of Pennsylvania. This means consistently updated data created by qualified groups of scholars.

William Bridges is the author of two bestselling books, *Transitions* and *Managing Transitions*. Bridges makes a distinction between change and transitions and states that it isn't the changes that do managers in; it is the transitions.

Bridges believes that change is situational, such as moving to a new home. He describes transition, on the other hand, as the psychological process people go through as they internalize and come to terms with the details of the new situation.

Bridges divides transition into three phases:

Ending phase
The neutral zone
The new beginning

Bridges asserts that if managers don't let go (ending phase) it is impossible to move through to the neutral zone. Moving into the neutral zone is necessary if managers want to reach the final stage, the new beginning. These stages have also been referred to as unfreezing, changing, and refreezing.

In more traditional textbooks, change management has three basic areas that should be examined: the actual task of managing change, the new body of knowledge that must be delivered, and how the change will impact the professional practice.

Change management requires political, analytical, people, business, and system skills. Organizations are social systems. Without people, there can be no organization. Guessing won't do. Change agents must learn to take apart and put together components, considering the financial and political impact.

There is no single change strategy. When developing a change strategy, it is important to consider that successful change is based on the communication of information. Redefining and reinterpreting existing norms and values and developing commitments to new ones is also essential for success.

The exercise of authority and the imposition of sanctions may be necessary but should be carefully considered before implementing. Every effort should be made to effectively build a new organization and gradually transfer people from the old one.

Other things to consider include the following:

Degree of resistance
Target population
Stakes
Time frame
Expertise

Dependency levels should be considered. Mutual dependency almost always signals the need for negotiation. If the organization is dependent on its people, management's ability to command or demand is limited. Conversely, if people are dependent upon the organization, their ability to oppose or resist is limited.

Managing change is more about leadership than management. A clear sense of mission or purpose is essential. Making that mission statement clear and simple is important.

Managing change also involves planning and organizing a sequence of activities that promote administrative and staff interaction. Change may involve policies, programs, organizational culture, physical environment, procedures, or relationships.

Change in organizations may lead to more efficient and cost-effective operations or could end in disaster.

Most change efforts require changes to organizational processes. The first step in this process is to truly understand the current processes. Managers of organizations today face a demand for change in their organizations if only because change is so pervasive in the world around them. Either we manage change, or we are managed by change.

There are several reasons for resistance to change from employees. These reasons include fear of the unknown, job security, bad timing, lack of resources, no personal gain, and fear of incompetence.

Change is rarely an easy process, especially within an organization. Poor-performing organizational cultures are generally those with the inability to change and adapt.

The most successful change agents manage the people side of change, not just the business side. Developing a change management strategy as well as the corresponding communication plan will actively manage much of the resistance to change.

It is important to note that the theories around change management can have a slightly different spin depending on the industry.

In the field of quality, the term change management refers to the processes by which new initiatives or systems are introduced and integrated into organizations. However, change management in the information technology world may lean more toward re-engineering and be about the correction or modification of old projects.

It is important to have conversations about change to ensure that everyone is using the same vocabulary and that they have similar expectations regarding the outcome. In the past, HR departments have acted as guides and facilitators for these conversations. Now Lean and Agile PMs are often required to take the helm.

As a Lean and Agile PM given the responsibility to facilitate a conversation on change and/or to simply deliver information regarding the change, keep in mind that there is a positive side to this exhausting process.

Whenever the Lean and Agile PM becomes the facilitator, it improves the credibility of the message. The facilitation process itself is a powerful method. It is a process that may be used to create detailed plans supported by the team. It can be used to explore options that may not have been considered.

Lean and Agile PMs might attend the session with participants, but they have a greater impact when they are actively involved in delivering the message.

Here are four quick facilitation tips:

Get the audience actively involved quickly.
Use visuals that have impact.
Simple is better than cluttered.
Don't assume everyone has the same level of knowledge about the
    situation.

A popular saying is "if change is the only constant, why are organizations so bad at it?" This is often meant to discourage Lean and Agile PMs. Most employees agree that change and success go hand in hand. Still, it is often said that no one likes change or that no one handles change well.

Statements that may be perceived as negative should be used to consider the fact that even positive change is something to be endured or tolerated.

However, there are people and organizations that do handle change better than others. Some even thrive on change. These entities appear to share some fundamental attributes. Generally, these organizations do the following:

They are not overextended.
They have formed friendly bonds within the company as well as the
    community.
They have made a practice of sharing information freely.
They are not militaristic in nature and have distributed the power.
They share a common purpose or story.

It is important to remember that, even though change disrupts teamwork, it is possible to build a successful team in the middle of transition.

Martin Luther King Jr., Margaret Thatcher, John F. Kennedy, Ronald Reagan, and Mother Teresa—with resolute courage and determination—stood squarely in the center of change and controversy. They were all successful in leading people with their thoughts and actions.

Probably the most confusing thing for employees relative to change is the fact that in today's workforce all employees are expected to exercise more choice and exercise more control over their time and work.

Often, changes made by the executive leadership team actually limit or take away that control. An employee must readjust even if he or she does not agree with or understand the change. This anxiety is lessened by better communication.

The phenomenon of changing minds is one of the least examined and least understood of human experiences. The ability to change a person's mindset involves the following:

Reason
Research
Resonance
Representation
Resources and rewards
Real-world events

Using reason on those individuals who consider themselves well-educated and research on those audiences who have the benefit of scientific training may be seen as manipulative. But using these mindsets, as long as they are not applied unethically, can help a Lean and Agile PM communicate the message more clearly.

For employees who need to have a feeling that the change is right, using resonance is more appealing on an emotional level.

Employees who are more visual may need representational stories that describe the event. This may include taking information and placing it in a different form, such as graphics or numbers.

Sometimes, resources and rewards or even the promise of these things can change a person's mind, and sometimes, real-world events, such as wars, hurricanes, or terrorist attacks, are what changes an employee's mindset.

Within the parameters of change management in business, there are four variables that should be assessed:

The magnitude of the proposed change
People's ability to make the change
People's willingness to make the change
The timing of the change

A popular scientific approach to change management is known as management transformation theory. This theory frequently references the following terms and attaches these meanings:

Management: the way we do the things we do
Management transformation: the way we improve management
Culture: the summation of individual community members' attitudes and behaviors
Cultural change: the metamorphosis of culture accomplished through the alteration of individual attitudes and behaviors

Management transformation is about improving management. This means continually critiquing and/or changing the way we do things. This can cause confusion or chaos as referenced earlier. Managers must learn to respect this condition as normal and treat it simply as a data point.

Change can be a difficult and costly problem, especially if that change also means changing requirements or a process. It is a worthwhile exercise to consider different approaches to change. There are three solid and recognized strategies for achieving the goal of minimizing the adverse impact of change. They include the following:

Identifying change early in the lifecycle
Facilitating the incorporation of change as soon as possible
Designing ways to reduce the impact of change

It is an essential skill for managers to not only manage their staff but also themselves during the change process. Often managers neglect or ignore their own feelings and emotions about the change taking place. A common reaction is anger. Anger is a common but destructive emotion in the workplace. Lean and Agile PMs may be in the position that they were not consulted about the change or may be asked to deliver a message they do not totally support.

It is important that Lean and Agile PMs explore methods for handling change, seeking an advisor if the reaction is anger. It is the manager's reaction more than the actual words that employees are likely to model.

Another reaction from Lean and Agile PMs to change can be that of apathy. If a Lean and Agile PM has been defeated on several business fronts, it is easy to get complacent and not present the change to the employees as important. Then, when changes are ignored, the staff is blamed for not rising to the occasion.

Most factors that influence change, such as globalization, technological advancement, re-engineering, and downsizing, are here to stay.

One common model used in describing the reaction to change follows a pattern similar to the reaction to the six stages of grief. Understanding these phases along with the reality that employees may move through these phases at a different pace can give managers insight regarding the change process.

Phase one: shock
Phase two: denial
Phase three: anger
Phase four: passive acceptance
Phase five: exploration
Phase six: challenge

Lean and Agile PMs must condition themselves to see these reactions more like data points. A common reaction may be to deny the impact of the change or react with an emotion, such as anger. A Lean and Agile PM, who does not realize an employee may still be reacting to phase one, shock, may not be as sensitive to the emotional outburst.

Most change management consultants will advise Lean and Agile PMs to expect chaos at first. Chaos obviously interferes with productivity. It takes a very small change to render a system chaotic. To avoid chaos, Lean and Agile PMs must be clear about their strategies, practices, policies, processes, behaviors, and expected outcomes.

Chaos and uncertainty can hinder the Lean and Agile PM's ability to achieve organizational goals. It can jeopardize the Lean and Agile PM's ability to implement long-lasting change strategies as well. Not all chaos is bad, and some chaos is necessary to develop creative energy.

Generally speaking, managers will want to minimize the chaos and help employees move peacefully toward change. Change management theory isn't based purely on the study of human and organizational behavior. It is a mixture of the following:

Psychology
Sociology
Business administration
Economics
Industrial and systems engineering.

Many projects using Six Sigma methodology face some resistance to change. Therefore, Lean and Agile PMs who are familiar with Six Sigma have an advantage when it comes to managing change. Basic Six Sigma principles support the theory that without user acceptance, any process improvement is doomed to fail.

Many aspects of change management are covered under the concept of organizational development (OD) covered in Chapter 9, Organizational Development and Performance Management Considerations for Lean and Agile Project Managers. Many professionals in the field of OD are continually reassessing through dialogue, debate, and discussions. Change management plays a crucial role in all OD efforts.

The need for change is increasing. Change capability is necessary for the organizations that will succeed in the future. There is a connection between change management and leadership development as well as basic employee education. Just as ethics is intertwined in all aspects of human resource management, so is the topic of change. People resist change in many ways. The most common include the following:

Ignoring the new process
Disagreeing
Criticizing the tools
Delaying the implementation

Making the process impossible to ignore can be accomplished by tying the success of the process to personal training plans or annual reviews. Evidence that other companies or departments have been successful with similar efforts can assist in gaining acceptance. It becomes difficult to argue against change when there is a proven track record in place by others. Providing a way for employees to constructively provide feedback will lessen criticism of the tools. Demonstrating the benefit of the change will encourage employees not to delay the implementation.

Changes can be caused by employee and management turnover, changes in market conditions, or economic fluctuations. Keeping people informed about upcoming change is critical. This avoids the surprise and fear that arises from uncertainty. It is important to pay close attention to communications. A positive spin will lessen employee anxiety. Examples of communications that should be carefully scrutinized include the following:

Newsletter articles
Organizational change announcements
Announcements for all-hands meetings
Announcements regarding promotions or employee terminations

Change management starts and ends with individuals. At the heart of modern organizations are teams that share the responsibility and the resources for getting things done. Change processes are mostly initiated by individuals or small teams, but the focus of change is one which goes beyond that small unit. As a Lean and Agile project manager it is important to understand when a major shift occurs. The most common error in managing change is underestimating the impact it has on employees.

Employees often associate change with a loss of security, relationships, and territories. A Lean and Agile PM must manage that sense of loss in order to lead the team in the new direction.

Ordering or demanding change rarely works. It is more common for employees to change because of encouragement and support. This can be accomplished by creating recognition for change agents and those who actively remove roadblocks to change including the following:

Ways to create recognition: The PM needs to establish criteria for what performance or contribution constitutes rewardable behavior or actions.
All employees must be eligible for the recognition.
The recognition must supply the employer and employee with specific information about what behaviors or actions are being rewarded and recognized.
Anyone who then performs at the level or standard stated in the criteria receives the reward.
The recognition should occur as close to the performance of the actions as possible, so the recognition reinforces the behavior the employer wants to encourage.

Chapter 18

Lean and Agile Project Management: International Influences

Applying Lean thinking and Agile techniques (Lean and Agile) to government implies a new attitude that examines complex bureaucratic systems with the intent of simplifying procedures and reducing waste. Several governmental agencies have discovered that the Lean approach has enabled them to make complicated processes function better, faster, and more cost-effectively.

The Environmental Protection Agency (EPA) is a strong advocate of Lean and Agile. The EPA recently published several successful case studies. The International City/County Management Association (ICMA) supports a program to assist local government organizations with implementing Lean. The American Society for Quality (ASQ) advocates the adoption of Lean and/or Six Sigma within the U.S. federal government. Several U.S. political figures have endorsed the Lean initiatives.

During the 81st General Assembly in the state of Iowa, legislation was passed that authorized the Department of Management to create the Office of Lean Enterprise. In the January 2012 Colorado state address, Governor John Hickenlooper remarked that almost every department had initiated a Lean program in order to identify waste and inefficiencies and create savings. Colorado House Bill 11-1212 was passed to integrate Lean and Agile principles. This bill promotes incorporating Lean practices as well as training state employees to be Lean experts within the state of Colorado.

Clearly, in the United States, the practice of Lean and Agile is becoming more and more popular. Examples of government agencies with active Lean departments include, but are not limited to, the following:

U.S. Department of Defense
U.S. Army
U.S. Department of Agriculture
U.S. Department of Housing and Urban Development
U.S. Nuclear Regulatory Commission

Lean and Agile practices are moving abroad. In Sweden, the Migration Agency is widely regarded as one of the most prominent public authorities to have adopted the Lean model. In 2009, the Singapore Housing and Development Board teams used Lean tools to provide award-winning citizen service.

Using Lean tools, government entities can expect to do the following:

Eliminate or dramatically reduce backlogs
Reduce lead times
Decrease the complexity of processes
Improve the consistency of reviews or inspections
Benefit from better staffing allocation

The challenge with implementing Lean and Agile is that most government departments are organized around functions rather than processes. In many cases, necessary resources are not located in the same building. Cross-training programs are generally not supported. Changing or modifying this dynamic alone drastically reduces waiting time, redundancy, and/or rework, but the concept is not always met with enthusiasm.

Lean and Agile supports the idea of creating work cells. A work cell is formed by placing all the necessary resources in one area. Work cells permit cross-training opportunities and reduce both rework and redundancy. Work cells can better manage the first in, first out (FIFO) process, generally increasing citizen satisfaction. Work cells are designed to improve process flow, eliminate waste, and promote standardization.

A basic premise of Lean thinking is to study the value of the work people do and directly connect it to the quality of service provided for the citizen. These activities may cause stress among employees who have not been enlightened by the merits of Lean. Employees may fear job loss or loss of control of their daily activities.

The following activities should be observed with a high level of sensitivity:

Core processes
Current systems managing these processes
People involved in these processes
Innovation possibilities

In the initial process, value stream mapping (VSM) can be extremely useful for governmental agencies. VSM refers to the activity of developing a visual representation of how a particular process, product, or service flows through the system.

VSM also identifies time frames, handoffs, and resources involved throughout the process. VSM, similar to flowcharting, has a set of symbols that represent various processes, materials, and information. However, unlike flowcharting, VSM symbols are not standardized, and there are several variations. New VSM symbols may be created when necessary, or verbiage may be placed inside a rectangular box to provide explanation regarding that step. Once the map is created, it is easier to identify areas of overt as well as hidden waste. Bottlenecks, redundancy, and rework are also more apparent.

In the beginning, another useful tool is Kaizen events, also known as rapid improvement events. The idea behind Kaizen events is to identify process improvements that can be implemented immediately. Kaizen events are designed to yield quick results. The ancillary benefit is this often increases employee buy-in and morale.

Kaizen events typically bring together a cross-functional team for three to five days to study a specific process. It is important that the members of this team have the ability to make decisions for their group because commitments are made during this session.

Kaizen events are conducted by a facilitator who walks the group through a model for process improvement. Often this model is plan-do-check-act (PDCA). Depending on the nature of the project, the Define, Measure, Analyze, Improve, and Control (DMAIC) model may be used. Proprietary models, such as select, clarify, organize, run, evaluate (SCORE™) may also be used to conduct the session. Additionally, there is the more traditional and simplified Kaizen approach that promotes the following:

Assessment
Planning

Implementation
Evaluation

In the assessment phase, the major goal is to determine the critical-to-quality (CTQ) factors. After a consensus is reached on the CTQ factors, the next step is to develop metrics. In the planning phase, the process improvement intervention is discussed. In the implementation phase, the process improvement is implemented and monitored. Finally, the evaluation phase measures the results based on the metrics developed during the assessment phase.

The success of any rapid improvement event depends on the following:

Teamwork
Personal discipline
Employee morale

In addition to rapid improvement events, another way to kick off a Lean and Agile program is by initiating a workplace organizational model, such as 5S. Similar to a VSM, the 5S model offers visual validation. Comparable to a Kaizen event, 5S can be completed in a relatively short period of time.

The 5S model uses a list of five Japanese words, which, translated roughly into English, start with the letter S: sort, set in order, shine, standardize, and sustain. The 5S model is also used to organize physical space in such diverse areas as health care, warehouses, and retail.

A new term, used more and more often in government services, is Lean IT. Although Lean principles are well-established and have broad applicability, the move to IT is still emerging. Lean IT will increase in use as more governments go online to deliver better services. Although many governments have already made the move to electronic files, the method used to manage these files often mimics manual systems. This makes retrieval of critical data difficult and cumbersome. Lean IT for government will allow these services to be more user-friendly and easier to audit.

In government services, the most challenging task is managing work in progress (WIP). There is a common belief that work received cannot be completed within a short time frame. This is often true because governmental systems are set up to collect data but often lack the discipline to act quickly on the data collected. One value of Lean is that, used properly, daily processes and activities are immediately identified in the value stream. Knowing how many permits are issued in a particular period or being able to calculate what is needed for tomorrow is the first step in process improvement.

Easy information-gathering tools are used that do not require a vast amount of training or instruction to be effective. Lean uses ordinary metrics to calculate results. When WIP is increased, productivity and quality generally decrease. The immediate goal becomes reducing WIP.

Most Lean and Agile projects share the same goals:

Increase citizen satisfaction
Optimize the value delivered to the public
Involve employees in the continual improvement effort
Develop consistent metrics that are clear and concise

Types of governmental projects that have benefited from implementing Lean and/or Six Sigma include improving the following:

Documentation management
File archiving
Inventory management
Payment process
Permit process
Security clearance

Lean and Agile starts with a vision. In the United States, the Lean and Agile leadership vision is usually to provide an efficient environment in which citizens are satisfied and employees are happy. Internationally, the happiness factor is often not regarded as an element, and citizen satisfaction is second to governmental control.

One common factor, however, with international Lean and Agile is the commitment necessary for upper management to motivate the workforce. Another common factor is that this cannot be achieved without some sort of map of the ongoing process. Nevertheless, in many cultures, attaining a map or verbal validation of the current process is nearly impossible.

Lean and Agile can benefit from methodology and tools normally associated with Six Sigma. For example, Lean and Agile favors the PDCA model for problem solving. Many problems in government are far too complex to benefit from this model.

Some governmental issues may need a more robust model, such as the DMAIC model or a Design for Six Sigma (DFSS) model used in Six Sigma and Lean Six Sigma programs. Six Sigma and Lean Six Sigma tools that effectively analyze root cause or performance capability may also be beneficial.

Lean and Agile, like Lean Six Sigma, take full advantage of other business management tools that include balanced scorecard, strengths, weaknesses, opportunities, threats (SWOT) analysis, and benchmarking theory.

The purpose of Lean and Agile is about contributing to overall citizen satisfaction. This is accomplished by optimizing value and by delivering services faster. Lean and Agile involves employees in the problem-solving process and uses performance metrics to measure success.

Colorado House Bill 11-1212 provides a solid explanation of Lean principles, which may be applied to any public sector entity. It states the following:

> Lean & Agile principles means a continuous and rapid process improvement of state government by eliminating a department's non value-added processes and resources, providing feedback on process improvements that have the purpose of increasing a department's efficiency and effectiveness, and measuring the outcomes of such improvements.

Internationally, as well as domestically, awareness of the government infrastructure is necessary before attempting to initiate a process improvement. The hierarchy, hiring policy, and labor responsibilities need to be considered as well. Paying attention to diversity and remembering Lean principles will ensure Lean and Agile success.

Some of the companies using Lean with their international client base as well as certain Agile practices include:

- John Deere
- Parker Hannifin
- Textron
- Illinois Tool Works
- Intel
- Caterpillar Inc.
- Kimberley-Clark Corporation
- Nike

In the book *A Leaner Approach to Government and Public Utilities*, several international governments are embracing the concept that Lean and Agile can be used to strengthen project management methodologies.

Chapter 19

ISO 13053 International Standards for Six Sigma

For the project manager interested in applying Lean thinking and Agile techniques (Lean and Agile) in the global environment, it is important to be aware of the international standards.

A significant international standard that the Lean and Agile project manager should be familiar with is ISO 13053, relating to Six Sigma. Although the name of the standard implies that it is for Six Sigma, in actuality, most of the tools to make the Define, Measure, Analyze, Improve, and Control (DMAIC) model are Lean and much of the core suggestions are very Agile in nature.

ISO is a global federation of national standards bodies. ISO publishes a number of standards. Standards are requirements and/or best practices involved in improving an organization. Currently, there are more than 300 standards available. Many standards offer organizations the ability to apply for ISO certification. Certification means that, according to an ISO auditor, the organization involved has met the requirements set forth in a specific standard. The work of preparing the standard is carried out through ISO technical committees. These committees include subject matter experts (SMEs) as well as ISO representatives. However, some standards are intended as guidelines and do not offer certification.

The best-known ISO standards in the United States belong to the ISO 9000 series. ISO 9001:2008 is the most commonly used standard in the United States. First published in 1987, ISO 9001 is the original management standard. This standard has been updated many times. The objective of the standard is to provide a framework to assess a company's ability to meet the needs of the customer. Simplified, this standard requires organizations to (1) identify their quality management system (QMS) and (2) continually improve the QMS process. Entities must be registered with ISO to qualify for ISO certification.

Generally, the process of registration involves these steps:

Application is made to a certified ISO registrar.
An assessment is made by the registrar involving two steps:
Readiness survey
Quality management system review
Registration may be granted, or the organization may be required to perform a series of tasks prior to registration.

Generally speaking, companies hire a consultant to prepare for registration. The consultant may be independent or be an employee of a certified ISO registrar. Internal audits may also be performed by a consultant prior to the official audit. Several companies elect to train a few employees to perform these mock audit activities either for the initial certification or for a recertification effort. Recertification time frames are prescribed by the specific standard but can also be contingent on how well compliance to the standard is being met.

Sometimes, ISO 9001:2008 is referred to as ISO 9000 because it is part of the ISO 9000 series. However, the document ISO 9000 is a supporting document related to the fundamentals and vocabulary of the standard. Prior to the year 2000, there were separate ISO 9000 standards that governed companies responsible for making products in contrast to companies that handled only the distribution of products. These two standards, ISO 9002 and ISO 9003, are no longer supported. ISO 9004 is a guidance document that helps explain the requirements of ISO 9001:2008. If a specific section of the standard does not apply, an organization may request exclusion. A new revision of Standard 9001 was released in 2015.

Many Lean and Agile project managers already work with ISO 9001:2008 because the standard is suitable for all sizes and types of organizations, including hospitals and the health care industry. The Lean and Agile project manager is able to reflect an accurate picture of an organization's current state as well as create viable measurement and tracking systems. These competencies are fundamental to the ISO certification process.

The primary goal of ISO 9001:2008 is to increase customer satisfaction. This is supported by better management controls and engaging in continuous process improvement. The Lean and Agile project manager is able to impact this initiative by eliminating errors, reducing waste, and providing sustainability models.

The second most recognized standard in the United States is ISO 14001. ISO 14001 focuses on how environmental issues are managed. First published in 1996, this standard supports the principles of ISO 9001:2008 and adds environmental considerations. Generally, there are four major stages to the certification process. These include the following:

Environmental review
Environmental policy creation
Documenting the environmental management system (EMS)
Audit and review

The Lean and Agile project manager interested in working with ISO 14001 should be familiar with the environmental efforts of the company as well as any compliance issues for that specific industry and/or governmental regulations.

The introduction of ISO 13053 for Six Sigma is an exciting development for the Lean and Agile project manager. Although the standard is specifically named Six Sigma, it contains many components typically associated with Lean manufacturing, continuous improvement (CI), and operational excellence (OE). For the Lean and Agile project manager working in ISO 9000 or ISO 14000 environments, the Six Sigma standard adds another layer of credibility to process improvement.

Many Six Sigma professionals rely on the American Society of Quality (ASQ) Six Sigma Black Belt Body of Knowledge (ASQ-SSBOK). This document provides an outline of topics that should be understood for the ASQ Six Sigma black belt certification exam. Many topics listed in the ASQ-SSBOK are the same as those covered in ISO 13053. However, differences also are evident. For example, ASQ-SSBOK supports more references to the history and value of Six Sigma, leadership, and the maturity of teams. ISO 13053 places more emphasis on tools, implementation, and the maturity of an organization.

ISO 13053 is divided into two standards: ISO 13053-1 and ISO 13053-2. ISO 13053-1 covers the DMAIC methodology. ISO 13053-2 covers tools used in the DMAIC process.

ISO 13053-1

This part of ISO 13053 records the best practices that should be followed in each of the phases of the DMAIC model. It makes management recommendations and gives an overall understanding of the roles and responsibilities in a Six Sigma project. In the typical Lean Six Sigma project, often the Lean and Agile project manager will need to assume several roles. Understanding how each role should function independently offers insight as well as a solid checklist.

Activities involved in a Six Sigma project are outlined in the standard as gathering data, extracting information from those data, designing a solution, and ensuring the desired results are obtained. ISO 13053-1 states that a reliable financial management model should be in place before beginning a process improvement.

In contrast, the ASQ-SSBOK specifically notes that Six Sigma project awareness should include an understanding of market share, margin, and revenue growth. Specific emphasis is placed on the following:

Net present value (NPV)
Return on investment (ROI)
Cost of quality (COQ)

ISO 13053-1 promotes a basic maturity model. Maturity models are popular in other process improvement programs as well. Maturity model levels are intended to be used as markers and milestones. These levels may also be used to monitor success and to build evaluation metrics. These levels of maturity are summarized as the following:

Level 1: The starting point
Level 2: Managed
Level 3: Defined
Level 4: Quantitatively managed
Level 5: Optimized

This particular model is familiar to students of capability maturity model integrated (CMMI). CMMI is a process improvement approach that is designed to improve enterprise-wide performance. CMMI is often used in defense contracts or software-related projects. According to the Software Engineering Institute (SEI), CMMI helps "integrate traditionally separate organizational functions, set process improvement goals and priorities, provide guidance for quality processes, and provide a point of reference for appraising current processes."

Voice of the customer (VOC) is emphasized in ISO 13053-1. The ASQ-SSBOK also includes references to VOC, defined as attention to customer feedback and understanding customer requirements. Customer requirements are also known as critical-to-quality (CTQ) factors. CTQ factors are more heavily stressed in Lean Six Sigma as compared to ISO 13053-1 or ASQ-SSBOK, thus exceeding the objectives of both documents. In LSS, the voice of the employee (VOE) as well as the voice of the business (VOB) and voice of the process (VOP) are considered along with VOC.

As with all documents related to Six Sigma or Lean Six Sigma, ISO 13053-1 fully explains the sigma statistic and normal distribution table by using the term defects per million opportunities (DPMO).

ISO 13053-1 discusses cost of poor quality (COPQ) and relies on the total quality management (TQM) definition of this term. COPQ is incurred by producing and fixing defects resulting from an internal or external failure. Lean Six Sigma confirms this definition but also includes what-if scenarios in the explanation of COPQ. For example, how much would it cost not to do something? How much revenue would be lost?

The ISO 13053-1 standard explains roles within a Six Sigma project and the basic responsibilities to the project. For each role, a competency model is included as well as training suggestions to achieve these designations. Roles outlined in this standard consist of the following:

Champion
Yellow belt
Green belt
Black belt
Master black belt
Deployment manager

These terms are universally accepted in Six Sigma, but Lean Six Sigma often includes the following roles and responsibilities:

White belt
Process owner
Sponsor

The ASQ-SSBOK does not go into detail about the various roles and responsibilities but does place a premium on things not included in the ISO 13053-1 document, such as team types. The team types include the following:

Formal
Informal
Virtual
Cross-functional
Self-directed

The ASQ-SSBOK includes sections on team facilitation, team dynamics, and team communication as well as time management of the team. This document also notes the various stages of a team that include the following:

1. Forming
2. Storming
3. Norming
4. Performing
5. Adjourning

Although information on teaming is not currently included in IS0 13053-1, the information provided by ASQ-SSBOK and team-building principles used in Lean Six Sigma practices may be considered supporting documentation for the roles and responsibilities and competency models outlined in ISO 13053-1.

ISO 13053-1 outlines how to prioritize projects and offers suggestions for project selection originally introduced by Edwards Deming that include considerations such as the following:

Are there measures?
Will the potential project improve customer satisfaction?
Is the potential project aligned to at least one of the business measures?

Project scoping, as well as process inputs and outputs, are discussed prior to introducing the DMAIC model. Project scoping and documentation of the scope are crucial activities for CI and OE initiatives as well. The Lean and Agile project manager can often help in the CI and OE effort by clarifying the scope.

The ISO 13053-1 standard documents the five phases of the DMAIC model. The phases are covered in more depth in Section II of this book, Lean Six Sigma Curriculum Development and Self-Study for the Global Professional. ISO 13053-1 captures essential information for each phase of the DMAIC.

ISO 13053-2

The primary purpose of 13053-2 is to introduce tools that will help execute the DMAIC process. The following tools are introduced along with individual fact sheets:

Affinity diagram
Brainstorming
Cause-and-effect diagram
Control charts
CTQ tree diagram
Data collection plan
Descriptive statistics
Design of experiment
Determination of sample
Failure mode and effects analysis (FMEA)
Gantt chart
Hypothesis testing
Indicators of key performance
Kano
Measurement systems
Monitoring/control plan
Normality testing
Prioritization matrix
Process mapping and process
Project charter
Project review
Quality function deployment—house of quality (QFD)
Responsible, accountable, consulted, and informed (RACI) matrix
Regression and correlation
Reliability
Return on investment (ROI), costs, and accountability

Services delivery
SIPOC
Value stream
Waste

ASQ-SSBOK and the Lean Six Sigma toolkit recognize the above-referenced tools. Students of OE or CI programs likewise use these tools. However, the ASQ-SSBOK does not offer specific instructions or fact sheets for these tools.

The Lean Six Sigma philosophy will consider the scalability of the tool when making a tool decision. For example, if the project is small, tools such as DOE, hypothesis testing, and QFD may not be useful. LSS theory also supports the thought that, if the tool is not necessary, it may be abandoned. LSS further believes it is acceptable to modify a tool or use a tool creatively. These concepts are not promoted in ISO 13053-2 or the ASQ-SSBOK.

The ASQ-SSBOK also covers design for Six Sigma (DFSS), which is not covered in ISO 13053-2 although the standard does list and explain several tools that may be applied to DFSS projects. Lean Six Sigma addresses DFSS but generally refers to it as design for Lean Six Sigma (DFLSS).

DFSS or DFLSS is a process methodology used when no existing process is in place. A popular DFSS model is define, measure, analyze, design, verify (DMADV). The first three phases of the model are the same as the DMAIC model, which is why it is a popular choice for the Lean and Agile project manager. The argument for DFSS is that some process improvements must be created from scratch and therefore require a design component.

ASQ-SSBOK places more emphasis on specific statistics and manual calculations. ISO 13053-2 promotes statistical thinking captured within the tools as opposed to individual statistical knowledge. Lean Six Sigma tends to slant toward the use of MS Excel–based statistical software with the intent of simplifying statistical concepts.

In summary, if the Lean and Agile project manager has not worked in an ISO environment, reviewing ISO 9001:2008 is essential. This standard is valuable to the Lean and Agile project manager even if certification is not the goal. ISO standards, in general, provide strong and defendable guidelines on what should be done to implement, monitor, and evaluate process improvements. This particular standard provides specific instructions for a successful QMS.

The introduction of ISO 13053 for Six Sigma provides a visual road map for the Lean and Agile project manager. It enhances the credibility of process improvement procedures and provides a common vocabulary and guidelines that may be implemented and understood worldwide. The Lean and Agile project manager should also review and understand the ASQ-SSBOK as well as various Lean Six Sigma toolkits. The concepts governing CI and OE are equally important to consider.

Finally, the Lean and Agile project manager should remember that all process improvement best practices, standards, theories, and methodologies share the respect and use of basic project management principles. One thing everyone in businesses associated with process improvement or project management should be aware of is that many standard methodologies are adopting best practices from other bodies of knowledge. We no longer live in a world afraid to blend ideas to form the best roadmap for success.

Chapter 20

The Difference between Lean and Agile

There is certainly a lot of synergy between Lean and Agile. This is especially true when the tools are examined. There is a lot of cross-over, but there are many significant differences.

The main difference, is of course, how projects are managed. In the Lean world projects still take a fairly traditional approach. As mentioned in earlier chapters Lean relies heavily on the plan-do-check-act model, which is systematic in nature, whereas many Agile teams rely on Scrum.

Lean thinking, which can be applied to any project, originated from Lean manufacturing. Lean manufacturing focused on eliminating waste within manufacturing processes to improve productivity, efficiency, and effectiveness. Lean is more interested in cost reduction and prioritizes customer satisfaction over everything.

Agile techniques originated and largely still lives in the software development arena. Whereas Agile does emphasize customer satisfaction, it is also concerned with the inner dynamics of the team.

Documentation theory and methods vary drastically, and in most cases Lean still depends heavily on Work Breakdown Structures where Agile does not.

Lean encourages teams to operate as a whole rather than compartmentalizing them in order to increase efficiency and improve productivity. Agile concentrates on defined pieces of the overall work product.

Lean is effective in making things work faster and eliminating waste. These principles are important to Agile as well, but the environment is less certain and doesn't always require continual documentation that things are moving in the right direction.

The successful project manager understands the differences and the synergies between Lean and Agile. Knowing when to apply which project management construct to which project is a valuable competency. The more a project manager studies different ways to manage projects the more equipped they will be in determining the best approach.

The most important thing to remember is that Lean was designed to reduce waste and improve operational efficiency. Agile was designed to execute tasks over a short time frame, often collaborating in real time with the customer. The reasoning behind including both in this book is to provide options for a project manager who wants to increase their value to the project and the team by using various tools from both methodologies.

Appendix A: Lean and Agile Project Management Body of Knowledge (SSD Global Solutions Version 4.2)

Over the past decade, traditional project management, Lean thinking, and Agile techniques have all evolved. To make things leaner, SSD Global Solutions (SSD) has labeled Lean and Agile project management as LAPM. We offer certifications that support this body of knowledge.

In this latest release, 4.2 LAPM has adopted many tools and ideologies that were not originally based on project management, Lean thinking, or Agile techniques. Part of the LAPM body of knowledge (BOK) includes highly respected business tools.

The LAPM BOK continues to improve and capitalize on thoughts that contribute to process improvement and project management. This forms a unique body of knowledge that borrows from a number of project management and process improvement theories.

The LAPM practice, itself, has become better, faster, and more cost-effective as a methodology. In its new form, it is intended to work in concert with traditional project management.

Although many Six Sigma and basic problem-solving methods dominate the themes presented in LAPM, there is a strong total quality management (TQM) influence. Therefore, this LAPM BOK is presented in three sections:

Major programs and established BOKs that contributed to Lean and Agile project management (Appendix A: Section 1)

Lean and Agile project management theory (Appendix A: Section 2)
Core tools and knowledge used in Lean and Agile project management
  (Appendix A: Section 3)

Section 1: Major Programs and Established BOKs that Contributed to Lean and Agile Project Management

The Primary Recognized Process Improvement Programs

Total Quality Management (TQM)

Total quality management (TQM) is the foundation of most process improvement programs. The core TQM strategy is to embed the awareness of quality throughout the entire organization. Both Six Sigma and Lean manufacturing/thinking promote concepts and tools first introduced by TQM. TQM also means continuously improving processes and products as well as reducing waste. This is why TQM aligns closely with Lean and Agile project management.

The major difference between Lean and Agile project management and TQM is that the tools used in Lean and Agile project management are updated and less labor-intensive. Generally, the mission, goals, and philosophy of TQM are also represented in Lean and Agile project management.

Many TQM ideas and problem-solving tools can be traced back to the early 1920s, when statistical theory was applied to product quality control. The concept of applying mathematical and statistical models to improve product quality was further developed in Japan in the 1940s. This effort was led by U.S. Americans such as Edwards Deming and Joseph Juran. Deming was responsible for popularizing the idea whereas Juran wrote much of the original literature.

Deming was a protégé of Dr. Walter Shewhart. Juran also studied with Shewhart. Shewhart is sometimes referred to as the father of statistical quality control. Shewhart's contribution to quality focuses on control charts, special/common cause variation, and analytical statistical studies. Shewhart's work also concentrates on statistical process control (SPC). Often SPC is studied as a subset of TQM. SPC studies various charts and graphs to determine and monitor process capability.

Beginning in the 1980s, a new phase of quality control and management began. The focus widened from quality of products to quality of all issues, including service opportunities, within an organization. It was determined that many of the same mathematical and statistical models used to identify, monitor, and evaluate the quality of products could also be applied in the service industry.

In 1988, a significant step in quality management was made when the Malcolm Baldrige Award was established by the President of the United States. This national award recognizes companies for their quality contributions. Malcolm Baldrige was responsible for bringing quality to the government during the Reagan administration. The Baldrige program's mission is to improve competitiveness and performance related to quality.

The Baldrige program was a direct result of the TQM movement and includes the following:

Raising the awareness of performance excellence
Providing organizational assessment tools and criteria
Educating business leaders
Recognizing national role models in quality

TQM is a set of management practices throughout an organization, geared to ensure that the organization consistently meets or exceeds customer requirements. In a TQM effort, all members of an organization participate in improving processes, products, and services. Quality initiatives are not limited to the quality department.

Modern definitions of TQM include phrases such as customer focus, the involvement of all employees, continuous improvement, and the integration of quality management into the total organization.

Basic TQM supports the following:

Line management ownership
Employee involvement and empowerment
Challenging quantified goals and benchmarking
Focus on processes and improvement plans
Specific incorporation in strategic planning
Recognition and celebration

TQM has adopted several documents that are also used in other process improvement efforts including the Lean and Agile project management program. Typically, these documents are identified by the following titles:

Deming's 14 Points
Deming's 7 Deadly Diseases
The Deming Cycle
Joseph Juran's Roadmap for Quality Leadership
The Triple Constraint Model

In general terms, TQM is a management approach to long-term success through customer satisfaction and is based on the participation of all members of an organization in improving processes, products, and services.

Lean and Agile

In 2001, 17 software developers met in Utah and published The Manifesto for Lean & Agile Software Development. The Lean and Agile movement was not any methodology but was intended to restore balance. Although originally intended for software development, Lean and Agile became popular for project management circa 2006. Agile's 12 principles are summarized as follows:

1. Customer satisfaction
2. Welcoming change requirements
3. Frequent delivery
4. Daily cooperation
5. Projects built around motivated individuals
6. Face-to-face conversation
7. Progress measurement
8. Sustainable development
9. Attention to technical excellence
10. Simplicity
11. Self-organizing teams
12. Frequent meetings to reassess

Quality focuses on specific tools and techniques, such as continuous integration, automated testing, test-driven development, and other practices. Compared to traditional project management, Lean and Agile targets complex systems.

One of the differences between Lean and Agile and Six Sigma is the approach to quality and testing. In the DMAIC model as well as project management, a Waterfall approach is taken; in Lean and Agile, an iterative approach is taken. In every iteration, a small part of the project is developed. Lean and Agile introduces a mindset as opposed to a methodology; the approach implies greater flexibility at any stage of project management development.

Lean and Agile promotes cross-functional teams, adaptive planning, early delivery, and continuous improvement.

International Standards Organization (ISO)

The International Standards Organization (ISO), founded in 1947, is an international standard-setting body composed of representatives from various national standards organizations. ISO has developed over 18,000 international standards, making it the largest standards-developing organization in the world. The ISO 9000 and ISO 14000 series are the most well-known. However, up to 1,100 new ISO standards are published every year.

The ISO 9000 family specifically addresses quality management. This means what the organization does to fulfill the following:

The customer's quality requirements
Applicable regulatory requirements
Enhance customer satisfaction
Achieve continual improvement of its performance in pursuit of these objectives

The ISO 14000 family addresses environmental management. This means what the organization does to do the following:

Minimize harmful effects on the environment caused by its activities
Achieve continual improvement of its environmental performance

To be certified in an ISO standard, these steps are necessary:

Locating and selecting a registrar; this is a company who is certified by ISO to make the initial assessment and provide suggestions for your ISO program.
Creating an application and conducting a document review.
Participating in an assessment.
Completing the ISO registration.
Participating in a recertification effort.

ISO recertification efforts include gathering the proper measurements and articulating these measurements as well as identifying future opportunities for process improvement. There is also a time factor involved. Therefore, Lean and Agile project management often plays a primary role in ISO recertification.

Capability Maturity Model Integrated (CMMI)

Capability maturity model integration (CMMI) is another popular process-improvement program. This integrated approach is intended to help an organization improve performance by recognizing certain levels of performance. CMMI can be used to guide process improvement across a project, a division, or an entire organization.

In CMMI models with a staged representation, there are five maturity levels designated by the numbers 1 through 5:

1. Initial
2. Managed
3. Defined
4. Quantitatively managed
5. Optimizing

CMMI was developed by the CMMI project, which was designed to improve the usability of maturity models by integrating many different models into one framework. The project consisted of members of industry, government, and the Carnegie Mellon Software Engineering Institute (SEI). The main sponsors included the Office of the Secretary of Defense (OSD) and the National Defense Industrial Association.

Each level in the CMMI process requires detailed information gathering and analysis. The significance of Lean and Agile project management in CMMI is that often to move up one level Lean and Agile project management practices need to be engaged.

Six Sigma

Defect Reduction

The Six Sigma problem-solving methodology is the most effective tool to quickly reduce and eliminate defects. It is a team-based methodology that works by systematically identifying and controlling the process variables that contribute to producing the defect or mistake.

DMAIC Model

Improvement of existing products or processes using the Six Sigma methodology is done in five steps:

Define
Measure
Analyze
Improve
Control

Define

The purpose of the Define phase is to make sure that everyone understands the project and the goals of the process improvement effort. The basic steps include the following:

Create a process improvement charter and process map.
Identify or define the problems in your process that must be solved in order to meet or exceed the customer's specifications or expectations.
Identify and quantify customer requirements.
Identify and quantify the process output and defects that fall short of these requirements and create a problem statement.
State the project goal, which also must be a clear and measurable goal, and include a time limit for the project's completion.
Determine the few vital factors that are critical to quality, which need to be measured, analyzed, improved, and controlled.

Measure

The purpose of the Measure phase is to get a strong as-is snapshot of how the process is currently behaving. The basic steps include the following:

Select the critical-to-quality characteristics in your process. These are the outputs of the given process that are important to the customer. How are you doing now?
Define what that process output should be, which is done by looking at the customer requirements and the project goal.
Define the defect for the process. Remember, a defect is an output that falls outside the limits of customer's requirements or expectations and must be measurable.
Find the inputs to the process that contribute to defects.
Define the exact dollar impact of eliminating the defects in terms of increased profitability and/or cost savings.
Measure the defects that affect the critical-to-quality characteristics as well as any related factors.

Incorporate measurement systems analysis—a method to make sure the defects are being measured properly.

Analyze

The purpose of the Analyze phase is to review the measurements and information from the previous phase and determine, based on that information, what three to five solutions might be appropriate to solve the problem or roll out the activity. Steps include the following:

Determine root cause.
Identify variations that could be reduced.
Determine if correlation exists.
Do what-if scenarios.
Determine the timeline and cost of solutions.
Determine the sustainability of the solution.

Improve

The purpose of the Improve phase is to choose a solution, implement the solution, and be able to definitively prove that a process improvement has been accomplished. This is done by comparing the as-is state (Measure) with conditions after the process improvement has been rolled out. Basic steps include the following:

Articulate the three to five possible solutions.
Gain consensus on the best solution.
Pilot.
Create an execution plan (project plan) if the solution is successful in the pilot.
Choose another one of the three to five solutions if the pilot is not successful.
Roll out.

Control

The purpose of the Control phase is to sustain the improvement. Basic steps include the following:

Clearly articulating the process improvement achieved
Creating a control plan to keep the process in place
Designing a transition plan for the new owner

DFSS Model

Design for Six Sigma, also known as design for Lean and Agile project management (DFSS or DFLSS), is applicable only in situations where a new product or service needs to be designed or redesigned from the very beginning. Many supporters of the DMAIC design believe that this is accomplished in the Analyze and Improve phases of the DMAIC model. However, supporters of DFSS believe a design component is necessary. Recently models based on the DMAIC thinking process that do not have a design component are also referred to as DFSS or DFLSS models.

Today, the most popular DFFS model is define, measure, analyze, design, verify (DMADV). The DMADV model contains the first three phases of the DMAIC model. The last two phases, Improve and Control, are replaced by design and verify.

Design

Design details optimize the design, and plan for design verification. This phase may require simulations.

Verify

Verify the design, set up pilot runs, implement the production process, and hand it over to the process owner(s).

Statistical Thinking

Both the DMAIC and DMADV model are based on statistical thinking. The following principles form the basis for statistical thinking:

All work occurs in a system of interconnected processes.
Inherent variation exists in all processes.
Reducing variation is the key to successfully improving a process.

Recognizing Individual Tasks within the Process and Assigning Major Causes of Variability

To successfully analyze a process using statistical process control, it is important to break things down into the smallest elements possible, accepting that all processes have inherent variability and that variability can be measured. Data are used to understand variability based on the type of variability. Deming used statistical quality control techniques to identify special and common cause conditions in which common cause was the result of systematic variability while special cause was erratic and unpredictable.

Common Cause

Common cause variability occurs naturally in every process. Common cause variation is fluctuation caused by unknown factors resulting in a steady but random distribution of output around the average of the data. Natural or random variation that is inherent in a process over time affects every outcome of the process. If a process is in control, it has only common cause variation and can be said to be predictable. Common cause variations are due to the system itself and are somewhat expected. Examples of common causes of variability are

Variation in the weight of an extruded textile or plastic tubing
Variation in moisture content of a resin
Particle size distribution in a powder
Poor training

Special Cause

Special cause variation is usually assigned to one of the following conditions:
   Variation in the process that is assignable to a specific cause or causes. For example, a variation arises because of special circumstances. Special cause variation is variation that may be assigned to a specific cause. Examples of special cause variation are

The first labels on a roll of self-adhesive labels are damaged, marred, or otherwise unusable.
The cartons near the door of a warehouse are exposed to rain and ruined.

Stabilize Processes

Traditional tools for process stabilization include process capability studies and control charts. The Six Sigma methodology supports the concept that a process may be improved by simply stabilizing the process. Making a process stable means to bring the process within the upper and lower specification limits and as close to the norm as possible.

Lean Manufacturing/Lean Thinking

Whereas the Six Sigma model concentrates on defect and mistake reduction, Lean manufacturing and Lean thinking (service-related) concentrate on

Waste reduction
Speed
Voice of the customer, employee, business, process

Waste Reduction

In Lean manufacturing/thinking, other terms for waste are nonvalue, non–value added, and the Japanese term muda. The misconception about the term is that when items are identified as waste it does not necessarily mean that the item will be reduced or eliminated. It simply means that it does not contribute directly to the process being studied. The reduction of waste concentrates on eight key areas: transportation, inventory, motion, waiting, over-processing, overproduction, defects, and skills.

Speed

All process improvement programs are concerned with delivering a product or service that is cost-effective and has maintained a high degree of quality. Speed is also important but not as apparent in other process improvement programs. Speed is highly recognized in Lean manufacturing/thinking. One avenue for speed is automation. The term automation, like the term waste, is often misunderstood. Automation simply means standardizing processes, which is also a goal of Six Sigma.

Lean supports many philosophies to include Just-in-Time (JIT). Individually, these efforts are sometimes known as concentration of assembly, Kanban cards, bar coding, visible record systems, production leveling, and work standardization.

Voice of the Customer, Employee, Business, Process

One of the unique things about the Lean methodology is an emphasis on how the customer, employee, business, and process are impacted by the process improvement. This is often referred to as VOC, VOE, VOB, and VOP.

Additional Methodologies and Bodies of Knowledge That Play a Role in Lean and Agile Project Management

The Quality Body of Knowledge (Q-BOK™) is a collection of outlines and documents maintained by the American Society of Quality (ASQ). These outlines are used for general information, reference, and to study for a variety of ASQ certifications. The Q-BOK contains a Six Sigma green belt body of knowledge and a black belt Six Sigma body of knowledge. ASQ was the first to establish an industry-recognized body of knowledge for Six Sigma. ASQ currently does not have a Lean and Agile project management body of knowledge. However, the Lean and Agile project management body of knowledge (SSD Global Version 3.0) contains much of the industry-accepted documentation on Six Sigma.

The Business Analysis Body of Knowledge (BABOK®) is maintained by the International Institute of Business Analysis. It supports six knowledge areas:

Business analysis planning and monitoring is concerned with which business analysis activities are needed. This includes identifying the stakeholders.

Elicitation is obtaining requirements from the stakeholders.

Requirements management and communication deals with changes to requirements as well as communication to stakeholders.

Enterprise analysis defines the business need and a solution scope.

Requirements analysis is the progressive elaboration of requirements into something that can be implemented.

Solution assessment and validation determines which solution is best, identifies any modifications that need to be made to the solution, and assesses whether the solution meets the business needs.

The BABOK® provides a framework that describes the areas of knowledge related to business analysis. The BABOK® is intended to describe and define business analysis as a discipline, rather than define the responsibilities of a person. The Guide to the Business Analysis Body of Knowledge is not really a methodology, which makes it easy to partner with Lean and Agile project management.

First published in 2005 by the International Institute of Business Analysis (IIBA), it was written to serve the project management community. The IIBA® has created the Certified Business Analysis Professional™ (CBAP®), a designation awarded to candidates who have successfully demonstrated their expertise in this field. This is done by detailing hands-on work experience in business analysis through the CBAP® application process and passing the IIBA® CBAP® examination.

The Project Management Body of Knowledge (PMBOK®) is maintained by the Project Management Institute (PMI). All process improvement programs recognize that basic project management must be in place before process improvement may begin. The PMBOK® supports nine knowledge areas:

Integration management
Scope management
Time management
Cost management
Quality management
Human resource management
Communications management
Risk management
Procurement management

The PMBOK® also promotes that the following phases are necessary for a successful project:

Initiating
Planning
Executing
Monitoring and controlling
Closing

Business process reengineering (BPR) is an approach intended to elevate the efficiency and effectiveness of an existing business process. BPR is also known as business process redesign, business transformation, and business process change management. BPR supports the following methodologies for process improvement:

Process Identification
Audit of the Current Situation Prototype
Test and Implement

Change management has a variety of meanings depending on the area. All areas of change management play a role in the new Lean and Agile project management. These areas include the following:

Project management refers to a project management process in which changes are formally introduced and approved.

Information technology service management (ITSM) is a discipline used by IT professionals.

People change management is a structured approach to changing individuals, teams, organizations, and societies.

Leadership development traditionally has focused on developing leadership ability. In a Lean and Agile project management organization, these methods are imperative to the success of Lean and Agile project management black belts and master black belts. Successful leadership development is generally linked to the following:

Individual's ability to learn

Quality and nature of the leadership development program

Genuine support for the leader's supervisor

Leaders play a key role in building a successful Lean and Agile project management organization. There are four main areas of responsibility:

Choosing the right projects

Choosing the right people

Following the right methodology

Clearly defining roles and responsibilities

Measurement Systems Analysis (MSA) is a science that considers selecting the right measurement. Studying the measurement interactions along with assessing the measurement device is also part of the mix. Are measures reliable and valid? What is the measurement uncertainty?

Statistics is the science of making effective use of numerical data relating to groups of individuals or experiments. Six Sigma and Lean have always included the field of statistics when measuring and analyzing data. The Lean and Agile project manager has to make these studies more digestible for the everyday person. A stronger emphasis is placed on choosing the right software and making sure that the statistic is valid.

Business finance plays a stronger role for the Lean and Agile project manager. The buy-in and continued support of a project cannot be based solely on statistical data. Choosing the right return-on-investment formula and being able to measure project success using financial terms have become essential.

As we move forward as Lean and Agile project managers, it is important to remember that Lean and Agile project management is not just a matter of blending two highly successful process methodologies but rather encompassing a collection of bodies of knowledge.

Organizational development is a body of knowledge and practice that enhances organizational performance and individual development. Today's organizations operate in a rapidly changing environment. One of the most important assets for an organization is the ability to manage change. Although there is not an industry-standard established document outlining the things necessary for successful organizational development, most professionals in this field rely on the works of William Bridges. Bridges is known as one of the foremost thinkers and speakers in the areas of change management and personal transition. Themes throughout Bridges' work encourage recognizing the various phases of change, the most popular being freezing, changing, and refreezing.

Section 2: Lean and Agile Project Management Theory

Lean government implies a new attitude that examines complex bureaucratic systems with the intent of simplifying procedures and reducing waste. Several government agencies have discovered that the Lean approach has enabled them to make complicated processes function better, faster, and more cost-effectively.

The Environmental Protection Agency (EPA) is a strong advocate of Lean government. The EPA recently published several successful case studies. The International City/County Management Association (ICMA) supports a program to assist local government organizations with implementing Lean. The American Society for Quality (ASQ) advocates the adoption of Lean and/or Six Sigma within the U.S. federal government. Several U.S. political figures have endorsed the Lean initiatives.

During the 81st General Assembly in the state of Iowa, legislation was passed that authorized the Department of Management to create the Office of Lean Enterprise. In the January 2012 Colorado State Address, Governor

John Hickenlooper remarked that almost every department had initiated a Lean program in order to identify waste/inefficiencies and create savings. Colorado House Bill 11-1212 was passed to integrate Lean government principles. This bill promotes incorporating Lean practices as well as training state employees to be Lean experts within the state of Colorado.

Clearly, in the United States the practice of Lean government is becoming more and more popular. Examples of government agencies with active Lean departments include, but are not limited to, the following:

U.S. Department of Defense
U.S. Army
U.S. Department of Agriculture
U.S. Department of Housing and Urban Development
U.S. Nuclear Regulatory Commission

Lean government practices are moving abroad. In Sweden, the Migration Agency is widely regarded as one of the most prominent public authorities to have adopted the Lean model. In 2009, the Singapore Housing and Development Board teams used Lean tools to provide award-winning citizen service.

Using Lean tools, government entities can expect to do the following:

Eliminate or dramatically reduce backlogs
Reduce lead times
Decrease the complexity of processes
Improve the consistency of reviews or inspections
Benefit from better staffing allocation

The challenge with implementing Lean government is that most government departments are organized around functions rather than processes. In many cases, necessary resources are not located in the same building. Cross-training programs are generally not supported. Changing or modifying this dynamic alone drastically reduces waiting time, redundancy, and/or rework, but the concept is not always met with enthusiasm.

Lean government supports the idea of creating work cells. A work cell is formed by placing all the necessary resources in one area. Work cells permit cross-training opportunities and reduce both rework and redundancy. Work cells can better manage the first in, first out (FIFO) process, generally increasing citizen satisfaction. Work cells are designed to improve process flow, eliminate waste, and promote standardization.

A basic premise of Lean thinking is to study the value of the work people do and directly connect it to the quality of service provided for the citizen. These activities may cause stress among employees who have not been enlightened by the merits of Lean. Employees may fear job loss or loss of control of their daily activities.

The following activities should be observed with a high level of sensitivity:

Core processes
Current systems managing these processes
People involved in these processes
Innovation possibilities

In the initial process, value stream mapping (VSM) can be extremely useful for governmental agencies. VSM refers to the activity of developing a visual representation of how a particular process, product, or service flows through the system.

VSM also identifies time frames, handoffs, and resources involved throughout the process. VSM, similar to flowcharting, has a set of symbols that represent various processes, materials, and information. However, unlike flowcharting, VSM symbols are not standardized, and there are several variations. New VSM symbols may be created when necessary, or verbiage may be placed inside a rectangular box to provide explanation regarding that step. Once the map is created, it is easier to identify areas of overt as well as hidden waste. Bottlenecks, redundancy, and rework are also more apparent.

In the beginning, another useful tool is kaizen events, also known as rapid improvement events. The idea behind kaizen events is to identify process improvements that can be implemented immediately. Kaizen events are designed to yield quick results. The ancillary benefit is that this often increases employee buy-in and morale.

Kaizen events typically bring together a cross-functional team for three to five days to study a specific process. It is important that the members of this team have the ability to make decisions for their group because commitments are made during this session.

Kaizen events are conducted by a facilitator who walks the group through a model for process improvement. Often this model is plan-do-check-act (PDCA).

Depending on the nature of the project, the Define, Measure, Analyze, Improve, and Control (DMAIC) model may be used. Proprietary models, such as select, clarify, organize, run, evaluate (SCORE™) may also be used to conduct the session. Additionally, there is the more traditional and simplified kaizen approach that promotes the following:

Assessment
Planning
Implementation
Evaluation

In the assessment phase, the major goal is to determine the critical-to-quality (CTQ) factors. After a consensus is reached on the CTQ factors, the next step is to develop metrics. In the planning phase, the process improvement intervention is discussed. In the implementation phase, the process improvement is implemented and monitored. Finally, the evaluation phase measures the results based on the metrics developed during the assessment phase.

The success of any rapid improvement event depends on the following:

Teamwork
Personal discipline
Employee morale

In addition to rapid improvement events, another way to kick off a Lean government program is by initiating a workplace organizational model such as the 5S. Similar to a VSM, the 5S model offers visual validation. Comparable to a kaizen event, 5S can be completed in a relatively short period of time.

The 5S model uses a list of five Japanese words, which, translated roughly into English, start with the letter S: sort, set in order, shine, standardize, and sustain. The 5S model is also used to organize physical space in such diverse areas as health care, warehouses, and retail.

A new term, used more and more often in government services, is Lean IT. Although Lean principles are well-established and have broad applicability, the move to IT is still emerging. Lean IT will increase in use as more governments go online to deliver better services. Although many governments have already made the move to electronic files, the method used to manage these files often mimics manual systems. This makes retrieval of critical data difficult and cumbersome. Lean IT for government will allow these services to be more user-friendly and easier to audit.

In government services, the most challenging task is managing work in progress (WIP). There is a common belief that work received cannot be completed within a short time frame. This is often true because governmental systems are set up to collect data but often lack the discipline to act quickly on the data collected. One value of Lean is that, used properly, daily processes and activities are immediately identified in the value stream. Knowing how many permits are issued in a particular period or being able to calculate a need for tomorrow is the first step in process improvement.

Easy information-gathering tools are used that do not require a vast amount of training or instruction to be effective. Lean uses ordinary metrics to calculate results. When WIP is increased, productivity and quality generally decrease. The immediate goal becomes reducing WIP.

Most Lean government projects share the same goals:

Increase citizen satisfaction
Optimize the value delivered to the public
Involve employees in the continual improvement effort
Develop consistent metrics that are clear and concise

Types of governmental projects that have benefited from implementing Lean and/or Six Sigma include improving the following:

Documentation management
File archiving
Inventory management
Payment process
Permit process
Security clearance

Lean government starts with a vision. In the United States, the Lean government leadership vision is usually to provide an efficient environment in which citizens are satisfied and employees are happy. Internationally, the happiness factor is often not regarded as an element, and citizen satisfaction is second to governmental control.

One common factor, however, with international Lean government is the commitment necessary for upper management to motivate the workforce.

Another common factor is that this cannot be achieved without some sort of map of the ongoing process. Nevertheless, in many cultures, attaining a map or verbal validation of the current process is nearly impossible.

Lean government can benefit from methodology and tools normally associated with Six Sigma. For example, Lean government favors the PDCA model for problem solving. Many problems in government are far too complex to benefit from this model.

Some governmental issues may need a more robust model such as the DMAIC model or a DFSS model used in Six Sigma and Lean Six Sigma programs. Six Sigma and Lean Six Sigma tools that effectively analyze root cause or performance capability may also be beneficial.

Lean government, like Lean Six Sigma, takes full advantage of other business management tools that include balanced scorecard; strengths, weaknesses, opportunities, threats (SWOT) analysis; and benchmarking theory.

The purpose of Lean government is about contributing to overall citizen satisfaction. This is accomplished by optimizing value and by delivering services faster. Lean government involves employees in the problem-solving process and uses performance metrics to measure success.

Colorado House Bill 11-1212 provides a solid explanation of Lean principles, which may be applied to any public sector entity. It states,

> Lean government principles means a continuous and rapid process improvement of state government by eliminating a department's non value-added processes and resources, providing feedback on process improvements that have the purpose of increasing a department's efficiency and effectiveness, and measuring the outcomes of such improvements.

Internationally, as well as domestically, awareness of the government infrastructure is necessary before attempting to initiate a process improvement. The hierarchy, hiring policy, and labor responsibilities need to be considered as well. Paying attention to diversity and remembering Lean principles will ensure Lean government success.

In order for Lean to function properly, it is important to pay attention to the PDCA and basic project management. For example, the project management activities, including initiation, planning, executing, and controlling, are necessary for Lean. See Figure A.1.

Planning can be the most time-consuming phase of the total project. Planning a project includes the following steps.

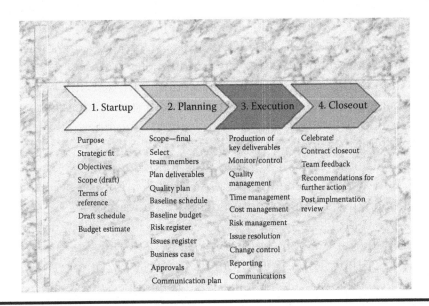

1. Startup	2. Planning	3. Execution	4. Closeout
Purpose	Scope—final	Production of key deliverables	Celebrate!
Strategic fit	Select team members	Monitor/control	Contract closeout
Objectives	Plan deliverables	Quality management	Team feedback
Scope (draft)	Quality plan		Recommendations for further action
Terms of reference	Baseline schedule	Time management	Post implmentation review
Draft schedule	Baseline budget	Cost management	
Budget estimate	Risk register	Risk management	
	Issues register	Issue resolution	
	Business case	Change control	
	Approvals	Reporting	
	Communication plan	Communications	

Figure A.1   Project life cycle.

1. Establish objectives.

   Your objective statement spells out the specific, quantifiable amount of improvement planned above the baseline performance that was indicated in the problem statement. You also need to determine how long completing this project and achieving your goal will take.

   The objective statement directly addresses the information in the problem statement. Just like the problem statement, the objective statement must contain certain information in order to be effective. A good objective statement contains all the following elements: metric, baseline, goal, amount of time, impact, and corporate goal/objective.

   That is, you want to improve some metric from some baseline to some goal in some amount of time with some impact against some corporate goal or objective. This timeline should be aggressive but realistic. These factors are necessary.

   Include the following elements in your objectives:

   Statement: A brief narrative description of what you want to achieve.

   Measures: Indicators you'll use to assess your achievement.

   Performance specifications: The value(s) of each measure that define success.

Lean and Agile project management still favors a popular method of setting SMART goals. SMART is an acronym that stands for specific, measurable, attainable, realistic, and timely.

To begin crafting your objective statement, start with the baseline performance you established in the problem statement. After you've set your improvement goal, you can estimate the financial benefit of achieving this goal.

Important questions to ask include the following:

Why? Why are we doing this project? Why is it important to the organization? Why is it important to me and the team?

What? What problems is the project expected to solve? What are the real issues at the core of the project? What deliverables do management, or the client expect from this project? What criteria will be used to judge success or failure? If we produce deliverables on time and on budget, what else represents success?

Who? Who has a stake in the outcome?

How? How do various stakeholders' goals differ?

The more clearly you define your project's objectives, the more likely you are to achieve them.

2. Develop a plan using work breakdown structure (WBS).

A work breakdown structure (WBS) is a breakdown of all the work important to finish a task. A WBS is orchestrated in a chain of importance and built to consider clear and coherent groupings, either by exercises or deliverables. The WBS should speak to the work distinguished in the affirmed project scope statement and serves as an early establishment for successful timetable advancement and expense evaluating. Supervisors commonly will build up a WBS as a forerunner to a nitty-gritty undertaking plan. The WBS should to be joined by a WBS dictionary, which records and characterizes WBS components.

The objectives of building up a WBS and WBS dictionary are (1) for the group to proactively and coherently arrange the task to fulfillment, (2) to gather the data about work that should be accomplished for an undertaking, and (3) to sort out exercises into sensible parts that will accomplish targets. The WBS and WBS dictionary are not the timetable but rather the building components. The movement of WBS and WBS dictionary advancement is as demonstrated in Figure A.2.

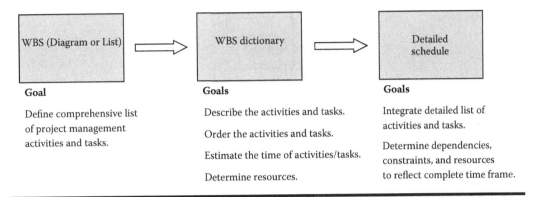

Figure A.2   WBS diagram.

The WBS and WBS dictionary are not static reports. WBS development is liable to administration dynamic elaboration, and as new data become known, the WBS should be overhauled to mirror that data. A project team that has significant changes to the WBS should reference the change management plan for direction on administration of changes to scope.

Example

Below is a simplified WBS example with a limited number of organizing levels. The following list describes key characteristics of a WBS (Figure A.3).

Hierarchical levels: Contains three levels of work.
Numbering sequence: Uses outline numbering as a unique identifier for all levels.
Level one is 1.0, which illustrates the project level.
Level two is 1.X (1.1, 1.2, 1.3, etc.), which is the summary level and often the level at which reporting is done.
Level three is 1.X.X (1.1.1, 1.1.2, etc.), which illustrates the work package level. The work package is the lowest level of the WBS at which both the cost and schedule can be reliably estimated.
Lowest level descriptions: Expressed using verbs and objects, such as "make menu."

WBS Numbering

In a WBS, each level has an allocated number with the goal that work can be recognized and followed after some time. A WBS may have shifting ideas

WBS example-banquet

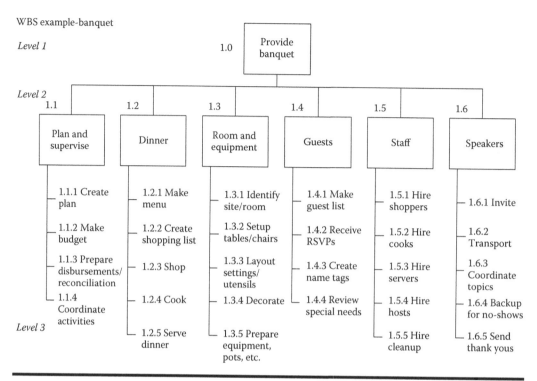

Figure A.3   WBS example.

for levels; however, there is a general plan for how to number each level. The following is the general tradition for how undertakings are decomposed:

> Level 1: Designated by 1.0. This level is the top level of the WBS and is usually the project name. All other levels are subordinate to this level.
> Level 2: Designated by 1.X (e.g., 1.1, 1.2). This level is the summary level.
> Level 3: Designated by 1.X.X (e.g., 1.1.1, 1.1.2). This third level comprises the subcomponents to each Level 2 summary element. This effort continues down until progressively subordinate levels are assigned for all work required for the entire project.

On the off-chance that assignments are legitimately subordinated, most booking apparatuses will consequently number errands utilizing the above tradition.

WBS Construction Methods

In spite of the fact that there are distinctive techniques for disintegrating work and making a WBS, the most direct and successful route is to utilize some type of visual showcase of the deliverables, stages, or exercises. In a perfect world, all project team individuals will gather and conceptualize all work required to finish deliverables effectively. There are two types of WBS:

Deliverable-oriented WBS
Process-centered WBS

Deliverable-Oriented WBS

A deliverable-oriented WBS is built around the project's desired outcomes or deliverables. This type of WBS would likely include the following characteristics:

Level 2 items are the names of all vendor project deliverables that are expected to be required as part of a contract. Level 2 should also include any agency deliverables tasks.

Level 3 items are key activities required to produce the Level 2 deliverables.

Additional levels are used depending upon the magnitude of the deliverables and the level of detail required to reliably estimate cost and schedule.

In the deliverable-oriented WBS, all deliverables are identified, and all work is included.

A procedure-focused WBS is like a deliverable-arranged WBS with the exception that it is sorted out, at the largest amount, by stages or in a procedure instead of by deliverables. The advantage of utilizing a procedure-focused WBS is that it empowers the incorporation of procedure-required deliverables, for example, project development life cycle (PDLC) deliverables. Notwithstanding the sort of WBS utilized, groups should guarantee that all legally binding and PDLC deliverables are represented in the WBS. A procedure-focused WBS ordinarily incorporates the accompanying items:

Level 2 exercises are stages or calendar checkpoints/turning points. These exercises could be PDLC stages, for example, initiation, planning, and so on.

Level 3 exercises are those exercises required to finish Level 2 stages or points of reference. Various undertakings are incorporated for any work that should be done in numerous stages.

Additional levels are utilized relying upon the length of the stage or plan and the level of subtle element required to dependably gauge cost and calendar.

In the procedure-focused WBS, all deliverables are recognized, and all work is incorporated. This completeness will diminish the danger of "reeling sheet" work undertakings, which may impact the plan.

Two industry-standard methods exist for determining how many levels a WBS should have:

Traditionally, the project management body of knowledge backs a foreordained seven-level model, which has the upside of clear names and meanings of every level (e.g., program, undertaking, subtask, work item, and level of exertion); the impediment to this model is that it requires a level of point of interest that might be superfluous. Models/strategies with foreordained levels and level definitions clarify what data should be incorporated and where, yet they need adaptability.

The more contemporary methodology is to give the attributes a chance to direct the quantity of levels utilized as a part of the judgment of the project manager. It is a decent practice to recognize the quantity of levels to be utilized and so keep up consistency when constructing the WBS. The quantity of levels must be adequate to permit the project manager to dependably gauge timetable and cost and successfully screen and control work bundles.

Example: WBS Dictionary

The project manager and team should talk about the WBS and decide the number of levels that is appropriate. The discussion should include the key points of interest. If any terms or conditions need clarification, a WBS dictionary should be attached. The terms placed in the WBS dictionary may be valuable later when creating benchmarks, determining the communication plan, or for group discussions.

WBS Dictionary—Table Format Example

WBS #:	1.1.1	Task:		Create plan	
Est. Level of Effort:	40 hrs.	Owner:		Project manager	
Resources Needed:	Subject matter experts	Work Products:		MS Project Plan	
Description of Task:	Development of a detailed project plan that lists all key resources, tasks, milestones, dependencies, and durations				
Input:	Approved project charter SMEs				
Dependencies:	Approval of budget				
Risk:	Changes to IT apps plans and deliverables IT apps implementation releases, which conflict with implementation				
WBS #:	1.1.2		Work Item:		Make budget
Est. Level of Effort:	16 hrs		Owner:		Project manager
Resources Needed:	CFO, CIO, executive sponsor		Work Products:		ITPR
Description of Task:	Development and documentation of the project budget based on plan and resources				
Input:	Approved project charter SMEs				
Dependencies:	Approval of project charter				
Risk:	Changes to IT apps plans and deliverables IT apps implementation releases which conflict with implementation				

WBS Fields					
WBS #	Task	Description of Task	Work Products	Owners	Est. Level of Effort
1	PLANNING	All task management and management activities			
1.1	Plan and Supervise		Roll-up task	Project manager	N/A
1.1.1	Create Plan	Development of WBS, work package identification, schedule formulation, staffing projection, resource estimation. Followed by development of a detailed project plan that lists all the key resources, task, milestones, dependencies, and duration.	WBS, WBS Dictionary, MS Project Plan	Project manager	40 hrs
1.1.2	Create Budget	Development and documentation of the project budget based on plan and resources.	ITPR	Project manager	40 hrs
1.1.3	Prepare Disbursement/ Reconciliation	Development of disbursement process for the project, including acceptance/ approval forms.	Purchase orders, deliverable product acceptance form	CFO	40 hrs
1.1.4	Coordinate Activities	Ongoing planning activities for the project including weekly meetings.	Meeting minutes	Project manager	8 hrs/week

The way to create a WBS and WBS dictionary is to engage in conversation about the activity and the steps necessary to achieve the undertaking. A project manager must guarantee that all the work that should be completed for the task is contained inside the WBS dictionary and is comprehended by colleagues and team members. A project manager should gather input from all team members to guarantee that the WBS and WBS dictionary are understandable and clearly identify the timing, cost, and resources by doing the following:

Scheduling a baseline
Determining a cost baseline
Scoping out a baseline
Quality baseline

Baselines are prepared on triple constraints—scope, time, cost (and quality). All of the above bullet points are considered as components of the project management plan. Often the scope, schedule, and cost baselines will be combined into one baseline that is used as an overall project baseline against which project performance can be measured. The performance measurement baseline is used for earned value measurements.

The outline refers to the estimated cost, resources needed (including labor costs), and the task schedule. Generally, the plan would include the following fields:

Original scheduled start and finish dates
Planned effort (may be expressed in hours)
Planned or budgeted cost
Planned or budgeted revenue

The main benefits of having a project baseline are the following:

Ability to assess performance
Earned value calculation
Improved future estimating accuracy

The job of the Lean and Agile project manager is to guide the team to successful delivery despite the challenges the world throws at the project. LAPM is about monitoring the project against the plan and intervening when the project manager notices things are going off track.

Section 3: Core Tools and Knowledge Used in Lean and Agile Project Management

Lean

Lean, Lean and Agile, and project management all have some common and agreed-upon tools.

Lean is a customer-centric methodology used to continuously improve any process through the elimination of waste in everything the project manager does; it is based on the ideas of continuous incremental improvement and respect for people.

Lean and Agile is a time-boxed, iterative approach to software delivery that builds software incrementally from the start of the project instead of trying to deliver it all at once near the end.

Project management is the application of processes, methods, knowledge, skills, and experience to achieve the project objectives. In general, a project is a unique, transient endeavor, undertaken to achieve planned objectives, which could be defined in terms of outputs, outcomes, or benefits.

The key tools in Lean include tools that make things work quickly and eliminate waste. They include the following:

5S

Organize the work area:

Sort (eliminate that which is not needed)
Set in order (organize remaining items)
Shine (clean and inspect work area)
Standardize (write standards for above)
Sustain (regularly apply the standards)

5S: The Purpose

Eliminates waste that results from a poorly organized work area (e.g., wasting time looking for a tool).

Andon

Andon is a visual feedback system for the plant floor that indicates production status alerts when assistance is needed and empowers operators to stop the production process.

Andon: The Purpose

Andon acts as a real-time communication tool for the plant floor that brings immediate attention to problems as they occur, so they can be instantly addressed.

Bottleneck Analysis

Bottleneck analysis identifies which part of the manufacturing process limits the overall throughput and improves the performance of that part of the process.

Bottleneck Analysis: The Purpose

Improves throughput by strengthening the weakest link in the manufacturing process.

Continuous Flow

Manufacturing in which work-in-process smoothly flows through production with minimal (or no) buffers between steps of the manufacturing process.

Continuous Flow: The Purpose

Eliminates many forms of waste (e.g., inventory, waiting time, and transport).

Gemba (The Real Place)

A philosophy that reminds project managers to get out of our offices and spend time on the plant floor—the place where real action occurs.

Gemba: The Purpose

Promotes a deep and thorough understanding of real-world manufacturing issues by first-hand observation and by talking with plant floor employees.

Heijunka (Level Scheduling)

A form of production scheduling that purposely manufactures in much smaller batches by sequencing (mixing) product variants within the same process.

Heijunka: The Purpose

Reduces lead times (because each product or variant is manufactured more frequently) and inventory (because batches are smaller).

Hoshin Kanri (Policy Deployment)

Align the goals of the company (strategy) with the plans of middle management (tactics) and the work performed on the plant floor (action).

Hoshin Kanri: The Purpose

Ensures that progress toward strategic goals is consistent and thorough, eliminating the waste that comes from poor communication and inconsistent direction.

Jidoka (Automation)

Design equipment to partially automate the manufacturing process (partial automation is typically much less expensive than full automation) and to automatically stop when defects are detected.

Jidoka: The Purpose

After Jidoka, workers can frequently monitor multiple stations (reducing labor costs), and many quality issues can be detected immediately (improving quality).

Just-in-Time (JIT)

Pull parts through production based on customer demand instead of pushing parts through production based on projected demand. Relies on many Lean tools, such as continuous flow, heijunka, Kanban, standardized work, and takt time.

Just-in-Time: The Purpose

Highly effective in reducing inventory levels. Improves cash flow and reduces space requirements.

Kaizen (Continuous Improvement)

A strategy with which employees work together proactively to achieve regular, incremental improvements in the manufacturing process.

Kaizen: The Purpose

Combines the collective talents of a company to create an engine for continually eliminating waste from manufacturing processes.

Kanban (Pull System)

A method of regulating the flow of goods both within the factory and with outside suppliers and customers. Based on automatic replenishment through signal cards that indicate when more goods are needed.

Kanban: The Purpose

Eliminates waste from inventory and overproduction. Can eliminate the need for physical inventories (instead relying on signal cards to indicate when more goods need to be ordered).

Key Performance Indicators (KPIs)

What are KPIs?

Metrics designed to track and encourage progress toward critical goals of the organization. Strongly promoted KPIs can be extremely powerful drivers of behavior, so it is important to carefully select KPIs that will drive desired behavior.

KPIs: The Purpose

The best manufacturing KPIs:

- Are aligned with top-level strategic goals (thus, the purpose is to achieve those goals).
- Are effective at exposing and quantifying waste (OEE is a good example).
- Are readily influenced by plant floor employees (so they can drive results).

Muda (Waste)
   Anything in the manufacturing process that does not add value from the customer's perspective.

Muda: The Purpose

There is none. Muda means waste. The elimination of muda (waste) is the primary focus of Lean manufacturing.

Overall Equipment Effectiveness (OEE)

Framework for measuring productivity loss for a given manufacturing process. Three categories of loss are tracked:

Availability (e.g., down time)
Performance (e.g., slow cycles)
Quality (e.g., rejects)

Overall Equipment Effectiveness: The Purpose

Provides a benchmark/baseline and a means to track progress in eliminating waste from a manufacturing process. One hundred percent OEE means perfect production (manufacturing only good parts, as fast as possible with no down time).

Plan-Do-Check-Act (PDCA)

An iterative methodology for implementing improvements:

Plan (establish plan and expected results).
Do (implement plan).
Check (verify expected results achieved).
Act (review and assess; do it again).

PDCA: The Purpose

Applies a scientific approach to making improvements:

Plan (develop a hypothesis).
Do (run experiment).
Check (evaluate results).
Act (refine your experiment; try again).

Poka-Yoke (Error-Proofing)

Design error detection and prevention into production processes with the goal of achieving zero defects.

Poka-Yoke: The Purpose

It is difficult (and expensive) to find all defects through inspection, and correcting defects typically gets significantly more expensive at each stage of production.

Root-Cause Analysis

A problem-solving methodology that focuses on resolving the underlying problem instead of applying quick fixes that only treat immediate symptoms of the problem. A common approach is to ask "why" five times, each time moving a step closer to discovering the true underlying problem.

Root-Cause Analysis: The Purpose

The purpose is to ensure that a problem is truly eliminated by applying corrective action to the root cause of the problem.

Single-Minute Exchange of Dies (SMED)

Reduce setup (changeover) time to less than 10 minutes. Techniques include the following:

Convert setup steps to be external (performed while the process is running).
Simplify internal setup (e.g., replace bolts with knobs and levers).
Eliminate nonessential operations.
Create standardized work instructions.

Single-Minute Exchange of Dies: The Purpose

Enables manufacturing in smaller lots, reduces inventory, and improves customer responsiveness.

Six Big Losses

Six categories of productivity loss that are almost universally experienced in manufacturing:

Breakdowns
Setup/adjustments

Small stops
Reduced speed
Startup rejects
Production rejects

Six Big Losses: The Purpose

Provides a framework for attacking the most common causes of waste in manufacturing.

SMART Goals

What are SMART Goals?

Goals that are specific, measurable, attainable, relevant, and time specific.

SMART Goals: The Purpose

The purpose is to ensure that goals are effective.

Standardized Work

Documented procedures for manufacturing that capture best practices (including the time to complete each task). Must be "living" documentation that is easy to change.

Standardized Work: The Purpose

Eliminates waste by consistently applying best practices. Forms a baseline for future improvement activities.

Takt Time

The pace of production (e.g., manufacturing one piece every 34 seconds) that aligns production with customer demand. Calculated as planned production time/customer demand.

Takt Time: The Purpose

Provides a simple, consistent, and intuitive method of pacing production. Is easily extended to provide an efficiency goal for the plant floor (actual pieces/target pieces).

Total Productive Maintenance (TPM)

A holistic approach to maintenance that focuses on proactive and preventative maintenance to maximize the operational time of equipment. TPM blurs the distinction between maintenance and production by placing a strong emphasis on empowering operators to maintain their equipment.

Total Productive Maintenance: The Purpose

Creates a shared responsibility for equipment that encourages greater involvement by plant floor workers. In the right environment, this can be very effective in improving productivity (increasing up time, reducing cycle times, and eliminating defects).

Value Stream Mapping

A tool used to visually map the flow of production. Shows the current and future state of processes in a way that highlights opportunities for improvement.

Value Stream Mapping: The Purpose

Exposes waste in the current processes and provides a roadmap for improvement through the future state.

Visual Factory

Visual indicators, displays, and controls used throughout manufacturing plants to improve communication of information.

Visual Factory: The Purpose

Makes the state and condition of manufacturing processes easily accessible and very clear to everyone.
The burn down chart is a fundamental metric in Lean and Agile.
The burn down chart is very simple. It is easy to explain and easy to understand. But there are pitfalls observed in many Lean and Agile workshops and adoptions.
People tend to think the burn down chart is so simple that they do not give appropriate attention to understand what it says.

Burn Down Chart

As a definition of this chart, the project manager can say that the burn down chart displays the remaining effort for a given period of time. When they track product development using the burn down chart, teams can use a sprint burn down chart and a release burn down chart.

Sprint Burn Down Chart

Teams use the sprint burn down chart to track the product development effort remaining in a sprint.

Generally speaking, the burn down chart should consist of the following:

X axis to display working days
Y axis to display remaining effort
Ideal effort as a guideline
Real progress of effort

Companies use different attributes on the Y axis. All of them have benefits and drawbacks.

Another popular tool in Lean and Agile is time boxing. In time management, time boxing allocates a fixed time period, called a time box, to each planned activity. Several project management approaches use time boxing. It is also used for project managers to address personal tasks in a smaller time frame.

In Lean and Agile, time boxing is a constraint used by teams to focus on value. One important time box that Lean and Agile promotes is the project itself. Contrary to Lean and Agile mythology, Lean and Agile teams prefer to have a time-boxed project because it offers a fixed schedule and a fixed team size.

Scrum meetings play an important role in Lean and Agile. Here is an overview of the different types of Scrum meetings:

1. Sprint planning meeting: This meeting begins with the product owner. This is when he or she explains the vision for the project as well as ways for the team to meet this goal. During this meeting, team members decide the amount of work they can complete in a timely manner. This is also when the team moves work from the product backlog to

the sprint backlog. This step requires a lot of planning, and usually it takes around eight hours for the group to decide on a finalized 30-day sprint.

2. Daily Scrum and sprint execution: From the planning meeting, the team moves into the daily Scrum meetings. Every single day for about 30 minutes, the team gathers together to report any issues or progress on their tasks. Although brief, this meeting is an essential part of the Scrum process. It is designed to keep all group members on track in a cohesive manner. Normally, the product owner is present during all daily Scrum meetings to assist in any way.

3. Sprint review meeting: This meeting is used to showcase a live demonstration of the work completed. During this meeting, the product owner, Scrum master, and stakeholders are present to review the product and suggest changes or improvements.

4. Sprint retrospective meeting: This meeting is held to facilitate a team's reflection on its progress. The team speaks openly about its organizational concerns and teamwork. During this meeting, dialogue should remain friendly, nonjudgmental, and impartial. This review session is a key part of team building and development, and it's also very important for future Scrum projects.

5. Backlog refinement meeting: The last type of Scrum meeting is the backlog refinement meeting. Team members focus on the quality and skill work involved during sprints. This meeting is necessary for the business owners to connect with the development team and is used to assess the quality and development of the final product. This meeting involves important reflection on the team backlogs. These backlogs are often written in user story form and specify what makes the product useful to the consumer.

Large or complex projects in big organizations often require some sort of executive "sponsorship" or leadership.

Any task that requires some preparation to achieve a successful outcome will probably be done better by using a few project management methods somewhere in the process. Project management methods can assist in the planning and managing of all sorts of tasks, especially complex activities.

Project management is chiefly associated with planning and managing change in an organization, but a project can also be something unrelated to business. Project management methods and tools can therefore be useful far more widely than people assume.

Project management involves the following:

Planning
Assessing/controlling risk
Allocation of resources
Organizing the work
Acquiring human and material resources
Assigning tasks
Tracking and reporting progress
Analyzing the results based on the facts achieved
Quality management
Solving issues

Typical types of documentation include the following:

Project charter
Work breakdown structure
Risk management plan
Communications plan
Project schedule
Stakeholder analysis

More important than any other topic in Lean and Agile project management is the project charter:

Should be the first step in all CI methodologies
Title/name of project
Project objectives
Scope
Assumptions and constraints
Cost factors and/or ROI
Cost of poor quality or cost of not doing the project

Appendix B: Lean and Agile Project Management Terms

4 Ms: Four-word categories used to provoke thought on an Ishikawa diagram (cause-and-effect diagram): material, method, machine, and man.

5 Whys: The practice of asking why five times when presented with a problem to try to identify potential root causes.

5S: A process and method for creating and maintaining an organized, clean, safe, and high-performing workplace. The steps are sort, set, shine, standardize, and sustain.

A3 Report: A Toyota-developed standard report showing a problem, analysis, and corrective action plan on a single piece of paper, usually A3 size.

A-B Control: A method used to regulate working relationships between a pair of operations such that overproduction is minimized. Machine A cannot feed machine B until it is empty or waiting for work.

Acceptance Criteria: Specific criteria identified by the customer for each functional requirement. The acceptance criteria are written in simple terms and from the perspective of the customer.

Acceptance Testing: Acceptance testing is a validation activity conducted to determine whether or not a system satisfies its acceptance criteria. It is the last phase of the software-testing process.

Affinity Diagram: Organizes brainstorming ideas into categories or themes. Useful when there are large amounts of information collected during a brainstorming session. It is also called the KJ method, after Kawakita Jiro (a Japanese anthropologist), who first developed the idea.

Agile: A conceptual framework for undertaking software projects. Agile methods are a family of development processes, not a single approach to software development.

Alpha Risk: The probability of accepting the alternate hypothesis when, in reality, the null hypothesis is true.

Alternative Hypothesis: A tentative explanation that indicates that an event does not follow a chance distribution—a contrast to the null hypothesis.

Andon: The Japanese word for a signal referring to a visual system that provides an indicator to supervision when abnormalities occur within processes.

ANOVA: Analysis of variance. This is a statistical test done by comparing the variances around the means of the condition being compared. In the simplest form, ANOVA provides a statistical test of whether the means of several groups are all equal.

ANOVA Gauge R&R: Measures the amount of variability induced in measure by the measurement system itself and compares it to the total variability observed to determine the viability of the measurement system.

Assignable Variation: Variation in data that can be attributed to specific causes.

Assumption: There may be external circumstances or events that must occur for the project to be successful (or that should happen to increase the chances of success). If it is believed that the probability of the event occurring is acceptable, it could be listed as an assumption. An assumption has a probability between 0% and 100%. That is, it is not impossible that the event will occur (0%) and it is not a fact (100%). It is somewhere in between. Assumptions are important because they set the context in which the entire remainder of the project is defined. If an assumption doesn't come through, the estimate and the rest of the project definition may no longer be valid.

Attribute: A characteristic that may take on only one value.

Attribute Data: Numerical information at the nominal level; subdivision is not conceptually meaningful data that represent the frequency of occurrence within some discrete category, for example, 42 solder shorts.

Automatic Line Stop: Ensuring that processes production will stop whenever a defect or problem occurs.

Autonomation: A term developed by Taiichi Ohno to describe "automation with human touch." These types of machines will stop when abnormalities occur so that they will not create large amounts of scrap and do not need an operator to watch the machine.

Average: Also called the mean, it is the arithmetic average of all the sample values. It is calculated by adding all of the sample values together and dividing by the number of elements (n) in the sample.

Background Variables: Variables that are of no experimental interest and are not held constant. The effects are often assumed insignificant or negligible, or they are randomized to ensure that contamination of the primary response does not occur.

Backlog: See Product Backlog.

Balance Chart: A bar chart or histogram that illustrates work content per operator. Can be used to balance work for operators or machines in order to achieve improvements in flow.

Balanced Scorecard: A performance management approach that focuses on customer perspective, internal business processes, and learning and growth and financials. It was originated by Dr. Robert Kaplan (Harvard Business School) and Dr. David Norton as a performance measurement framework that added strategic nonfinancial performance measures to traditional financial metrics to give managers and executives a more "balanced" view of organizational performance.

Batch and Queue: Typical mass production method such that a part going through a system will be produced in large batches to maximize "efficiency" and then sit in a queue waiting for the next operation.

Behavior-Driven Development: Behavior-driven development (or BDD) is an agile software development technique that encourages collaboration between developers, QA, and nontechnical or business participants in a software project. BDD focuses on obtaining a clear understanding of desired software behavior through discussion with stakeholders. It extends TDD by writing test cases in a natural language that nonprogrammers can read.

Benchmarking: A standard used to compare performance against best-in-class companies. It then uses the information gathered to improve its own performance. Subjects that can be benchmarked include strategies, products, programs, services, operations, processes, and procedures.

Beta Risk: The probability of accepting the null hypothesis when, in reality, the alternate hypothesis is true.

Blocking Variables: A relatively homogenous set of conditions within which different conditions of the primary variables are compared. Used to ensure that background variables do not contaminate the evaluation of a primary variable.

Bottleneck: A bottleneck is a sort of congestion in a system that occurs when workload arrives at a given point more quickly than that point can handle it. It is metaphorically derived from the flowing of water through a narrow-mouthed bottle in which the flow of water is constrained by the size of its neck.

Breakthrough Improvement: A rate of improvement at or near 70% over baseline performance of the as-is process characteristics.

Brownfield: A brownfield site is an existing facility that is usually managed in line with mass production methods.

Bugs: A software bug is a problem causing a program to crash or produce invalid output. It is caused by insufficient or erroneous logic and can be an error, mistake, defect, or fault.

Build to Order: A production environment in which a product or service can be made and assembled after receipt of a customer's order.

Burn Down Chart: A burn down chart is a visual tool for measuring and displaying progress. Visually, a burn down chart is simply a line chart representing remaining work overtime. Burn down charts are used to measure the progress of an Agile project at both an iteration and project level.

Capability: A measurement index that expresses the capability of the process by using a percentage.

Capital Linearity: A philosophy linked to capital expenditure on machinery such that a small amount of additional capacity can be added by using a number of smaller machines rather than one big and very expensive machine.

Casualty: The principle that every change implies the operation of a cause.

Causative: Effective as a cause.

Cause: That which produces an effect or brings about change.

Cause-and-Effect Diagram: This is also called a fishbone diagram. This is a graph that places the issue being discussed in the head of the fish. The bones of the fish are categories of problems that could be a problem. The smaller bones are the possible root causes.

Cell: A cell is a group of people, machines, materials, and methods arranged so that processing steps are located adjacent to each other and in sequential order. This allows parts to be processed one at a time or, in some cases, in a constant small batch that is maintained through the process sequence. The purpose of a cell is to achieve and maintain an efficient, continuous flow of work.

Center Line: The line on a statistical process control chart that represents the characteristic's central tendency.

Central Tendency: Data clustered around the middle. Mean, mode, and median are all examples of central tendency; numerical average, for example, mean, median, and mode; center line on a statistical process.

Chaku-Chaku: One-piece flow ideal whereby machines automatically unload parts so that an operator can move apart from one machine to the next without stopping to unload parts.

Champion: A person who supports the successful completion of the project.

Characteristic: A process input or output that can be measured and monitored.

Chief Engineer: The Toyota term used to describe the person who is totally responsible for the successful development of a product line.

Classification: Differentiation of variables.

Client/Customer: The person or group that is the direct beneficiary of a project or service is the client/customer. These are the people for whom the project is being undertaken (indirect beneficiaries are stakeholders). In many organizations, internal beneficiaries are called "clients" and external beneficiaries are called "customers," but this is not a hard and fast rule.

Common Cause: See Random Cause.

Common Causes of Variation: Sources of variability in a process that are truly random. These are generally inherent in the process itself and can be managed. This type of variation is usual, historical, and a quantifiable variation in a system.

Complexity: The level of difficulty to build, solve, or understand something based on the number of inputs, interactions, and uncertainty involved.

Confidence Level: The probability that a random variable x lies within a defined interval.

Confidence Limits: The two values that define the confidence level.

Confounding: Allowing two or more variables to vary together so that it is impossible to separate their unique effects.

Constraints: Constraints are limitations that are outside the control of the project team and need to be managed around. They are not necessarily problems. However, the project manager should be aware of constraints because they represent limitations that the project must execute within. Date constraints, for instance, imply that certain events (perhaps the end of the project) must occur by certain dates. Resources are almost always a constraint because they are not available in an unlimited supply.

Consumer Risk: The probability of accepting a lot when, in fact, the lot should have been rejected (see Beta Risk).

Continuous Data: A set of observations usually associated with physical measurement that can take on any mathematical value within specified parameters.

Continuous Flow: Each process, whether in an office or plant setting, makes or completes only the one piece that the next process needs; the batch size is one. Single-piece flow, or one-piece flow, is the opposite of a batch-and-queue process.

Continuous Random Variable: A random variable that can assume any value continuously in some specified variable.

Control Chart: The most powerful tool of statistical process control. It consists of a run chart, together with statistically determined upper and lower control limits and a centerline.

Control Limits: Upper and lower bounds in a control chart that are determined by the process itself. They can be used to detect special or common causes of variation.

Control Specifications: Specifications called for by the product being manufactured.

Cost of Poor Quality (COPQ): The costs associated with any activity that is not doing the right thing right the first time.

Critical Path: The series of consecutive activities that represent the longest time path through the process. The critical path is the sequence of activities that must be completed on schedule for the entire project to be completed on schedule. It is the longest duration path through the work plan. If an activity on the critical path is delayed by one day, the entire project will be delayed by one day (unless another activity on the critical path can be accelerated by one day).

Critical to Quality (CTQ): Any activity or thought related to the successful outcomes of the project.

Cross Dock: A facility that gathers and recombines a variety of inbound materials and parts from multiple suppliers to forward on to multiple customers.

Cutoff Point: The point that partitions the acceptance region from the reject region.

Cycle Efficiency (CE): CE is a measure of the relative efficiency in a production system. It represents the percentage of value-added time of a product through the critical path versus the total cycle time (TCT).

Cycle Time: The time a person needs to complete an assigned task or activity before starting again.

Cycle Time Interval: The frequency with which a particular item is made during a set period of time (usually days).

Daily Standup/Scrum: A daily standup is a whole team meeting that happens at the same time every day and usually lasts 15 minutes or less. The meeting is designed to allow the entire team to synchronize with each other and to understand the flow and challenges of the development process. Each team member should provide the following information What did I do yesterday, what am I planning to do today, and what impediments do I currently have?

Data: Factual information used as a basis for reasoning, discussion, or calculation; often refers to quantitative information.

Defect: An output of a process that does not meet a defined specification, requirement, or desire such as time, length, color, finish, quantity, temperature, etc.

Defective: A unit of product or service that contains at least one defect.

Degrees of Freedom: The number of independent measurements available for estimating a population parameter.

Deliverable: A deliverable is any tangible outcome that is produced by the project. All projects create deliverables. These can be documents, plans, computer systems, buildings, aircraft, etc. Internal deliverables are produced as a consequence of executing the project and are usually needed only by the project team. External deliverables are those that are created for clients and stakeholders. The project may create one or many deliverables.

Demand: The usage of an item over a period of time. This also includes an understanding of the customer requirements for quality, lead time, and price.

Density Function: The function that yields the probability that a particular random variable takes on any one of its possible values.

Dependent Variable: A response variable, for example, Y is the dependent or "response" variable where $Y = f(XI \ldots Xn)$ variable.

Design for Six Sigma (DFSS): The use of Six Sigma thinking, tools, and methods applied to the design of products. Any Six Sigma model for managing a project that is not DMAIC is generally considered a DFSS.

Design of Experiments (DOE): An efficient, structured, and proven approach to interrogating a process or system for the purpose of maximizing the gain in process or system knowledge.

Discrete Random Variable: A random variable that can assume values only from a definite number of discrete variables.

Distributions: The tendency of large numbers of observations to group themselves around some central value with a certain amount of variation or "scatter" on either side.

Done: Also referred to as "Done Done," this term is used to describe all the various tasks that need to happen before a story is considered potentially releasable.

DMAIC: This acronym stands for Define, Measure, Analyze, Improve, and Control. It is the heart of the Six Sigma process and refers to a data-driven quality strategy for improving processes. It is an integral part of any company's Six Sigma quality initiatives.

DPMO: Defects per million opportunities. The total number of defects observed divided by the total number of opportunities, expressed in parts per million.

DPU: Defects per unit. The total number of defects detected in some number of units divided by the total number of those units.

Effect: That which was produced by a cause.

Engineer to Order: Products whose customer's specifications are unique for each order; therefore, each product is engineered from scratch upon receipt of an order.

Epic: A very large user story that is eventually broken down into smaller stories.

Estimation: The process of agreeing on a size measurement for the stories as well as the tasks required to implement those stories in a product backlog.

Exits: The amount of work completed over a given amount of time measured in dollars or units.

Experiment: A test under defined conditions to determine an unknown effect, to illustrate or verify a known law, test or establish a hypothesis.

Experimental Error: Variation in observation made under identical test conditions, also called residual error. The amount of variation that cannot be the variables included in the experiment.

Factors: Independent variables.

Failure Mode and Effects Analysis (FMEA): A procedure used to identify, assess, and mitigate risks associated with potential product, system, or process failure modes.

Feature Creep: Feature creep occurs when software becomes complicated and difficult to use as a result of too many features.

FIFO: A strategically sized inventory that keeps the sequence of the production uniform throughout the value stream, maintaining flow.

Finish to Order: An environment such that products are built to as high a level as is possible and then configured to customers' requirements upon receipt of order.

Fishbone Diagram: See Cause-and-Effect Diagram.

Fixed Effects Model: Experimental treatments are specifically selected by the researcher. Conclusion only applies to the factor levels considered in the analysis. Inferences are restricted to the experimental levels.

Fixed-Position Stop System: A problem-addressing method on continuously moving production lines such that if a problem is identified and not resolved before a fixed point, the production line will stop.

Flowchart: A graphic model of the flow of activities, material, and/or information that occurs during a process.

Fluctuations: Variances in data that are caused by a large number of minute variations or differences.

Frequency Distribution: The pattern or shape formed by the group of measurements in a distribution.

Functional Manager: The functional manager is the person reported to within the functional organization. Typically, this is the person who does performance review. The project manager may also be a functional manager, but he or she does not have to be. If the project manager is different from the functional manager, the organization is probably utilizing matrix management.

Gage R&R: This is used in measurement systems analysis (MSA). A quantitative assessment of how much variation (repeatability and reproducibility) is in a measurement system compared to the total variation of the process or system.

Gantt Chart: A Gantt chart is a bar chart that depicts activities as blocks over time. The beginning and end of the block correspond to the beginning and end date of the activity.

Gemba: The Japanese term used to describe the "actual place" where value is added on the shop floor.

Greenfield: A new production facility not restricted by practices of the past; therefore, it has a culture of adapting to change without resistance.

Heijunka: Leveling the production by product and/or quantity over a fixed time period.

High-Level Value Stream Map: A visual representation of the aggregated material and information flows within a company or business unit.

Histogram: A bar chart that depicts the frequencies (by the height of the plotted bars) of numerical or measurement categories.

Homogeneity of Variance: The variances of the groups being contrasted are equal (as defined by statistical test of significant differences).

Hoshin: The Japanese word for planning; it is used throughout operational, financial, strategic, and project-based scenarios.

Independent Variable: A controlled variable; a variable whose value is independent of the value of another variable.

Input: A resource consumed, utilized, or added to a process or system. Synonymous with X, characteristic, and input variable.

Inspection: Mass production would use inspectors outside of a process. Lean producers assign the responsibility of quality to the areas in which the processes are performed. Inspections are performed within the areas that own the assembly process.

Instability: Unnaturally large fluctuations in a pattern.

Interaction: The combined effect of two factors observed over and above the singular effect of each factor against the level of the other factor. A significant interaction indicates that the effect of each factor on the response changes depend on the value of the other factor.

Interval: Numeric categories with equal units of measure but no absolute zero point, that is, quality scale or index.

Inventory Turns: A measure to quantify the pace at which inventory rotates throughout a company. Inventory turns = annual cost of goods sold/average value of inventory during year.

Issue: An issue is a major problem that will impede the progress of the project and that can't be resolved by the project manager and project team without outside help. Project managers should proactively deal with issues through a defined issues management process.

Jidoka: Quality built into processes such that if a process is not capable of creating the required output then it will not operate until it can.

Jishuken: A Japanese word used to describe a "hands-on learning workshop."

JIT: Stands for "Just in Time." This means producing or conveying only the items that are needed by the next process when they are needed and in the quantity needed. This process can even be used between facilities or companies.

Kaikaku: Radical improvement designed to quickly eliminate and/or add value to a value stream. Also described as breakthrough kaizen.

Kaizen: An incremental change for the better. The organized use of common sense to improve cost, quality, delivery, safety, and responsiveness to customer needs.

Kaizen Event: A rapid improvement event.

Kanban: Kanban, pronounced /ˈkɑnˈbɑn/, is a method for developing products with an emphasis on Just-in-Time delivery and the optimization of flow of work on the team. It emphasizes that developers pull work from a queue, and the process, from definition of a task to its delivery to the customer, is displayed for participants to see.

Kanban Board: Is also called a Scrum board displaying a sticky note for each task in progress. These are aligned in separate columns based on their status. The status on the board has three categories: to-do, doing, done.

Kanban Post: A storage container for Kanban cards pulling deliveries.

Labor Linearity: A manning philosophy such that as demand increases or reduces, manpower is added one at a time as such manpower requirements are linear to production volume.

Lead Time: The total time from the beginning of the supply chain to the time something needs to ship. The sum of the VA/NVA time for a product to move through the entire value stream.

Lean: Lean software development is a translation of Lean manufacturing and Lean IT principles and practices to the software development domain. Adapted from the Toyota production system and is a set of techniques and principles for delivering more values with the same or fewer resources by eliminating waste across organizations and business processes.

Lean Transactional: The application of Lean to business processes, such as paperwork flow through an office in accounts or marketing.

Level Selling: The eliminating of sales spikes generated by end-of-month sales targets at dealers and so forth. This allows for improved flow of demand from the customer and improvements in anticipated demand.

Life Cycle: Life cycle refers to the process used to build the deliverables produced by the project. There are many models for a project life cycle. For software development, the entire life cycle might consist of planning, analysis, design, construct/test, implementation, and support. This is an example of a "Waterfall" life cycle. Other life cycles include iterative development, package implementation, and research and development. Each of these life cycle models represents an approach to building the deliverables on the project.

Line Charts: Charts used to track the performance without relationship to process capability of control limits.

Linear Regression: Analyzes the relationship between two variables, X and Y.

Long-Term Variation: The observed variation of an input or output characteristic that has had the opportunity to be observed over time.

Lower Control Limit (LCL): Used in control charts to show the lower limit. Typically, three standard deviations below the central tendency.

Machine Cycle Time: The amount of time the unit spends in the operational cycle of a machine.

Mean: The statistical measure on a sample that is used as an estimate of the mean of the population from which the sample was drawn. Numerically, it equals the sum of scores divided by the number of samples.

Measurement Accuracy: For a repeated measurement, it is a comparison of the average of the measurements compared to some known standard.

Measurement Precision: For a repeated measurement, it is the amount of variation that exists in the measured values.

Median: The middle value of a data set when the values are arranged in either ascending or descending order.

Metric: A measure that is considered to be a key indicator of performance. It should be linked to goals or objectives and carefully monitored.

Milestone: A milestone is a scheduling event that signifies the completion of a major deliverable or a set of related deliverables. A milestone, by definition, has duration of zero and no effort. There is no work associated with a milestone. It is a flag in the work plan to signify that some other work has been completed. Usually, a milestone is used as a project checkpoint to validate how the project is progressing. In many cases, there is a decision, such as validating that the project is ready to proceed further, that needs to be made at a milestone.

Milk Run: Reducing transport costs and batch sizes by performing multiple pick up and drops at multiple suppliers using one truck.

Mixed Effects Model: Contain elements of both the fixed and random effects models.

Muda: The Japanese word for waste or non–value added activity.

Mura: The Japanese word used to describe variation or fluctuation.

Muri: The Japanese word used to describe overburdening or strain/stress.

Nemawashi: A Japanese expression used to describe the practice of obtaining support and buy-in for change by presenting the idea and then planning with upper management and stakeholders. Directly translated, it means "preparing the ground for planting."

Nominal: Unordered categories that indicate membership or non-membership with no implication of quantity, that is, assembly area number one, part numbers, etc.

Nonconforming Unit: A unit that does not conform to one or more specifications, standards, and/or requirements.

Nonconformity: A condition within a unit that does not conform to some specific specification, standard, and/or requirement, often referred to as a defect. Any given nonconforming unit can have the potential for more than one nonconformity.

Non–Value Added (NVA): Any activity performed in producing a product or delivering a service that does not add value.

Normal Distribution: The distribution characterized by the smooth, bell-shaped curve.

Null Hypothesis: A tentative explanation that indicates that a chance distribution is operating.

Obeya: Translated as "big room." This is the expression used by the Japanese to describe the powerful project room concept also known as a "war room."

Objective: An objective is a concrete statement that describes what the project is trying to achieve. The objective should be written at a low level, so that it can be evaluated at the conclusion of a project to see whether it was achieved. Project success is determined based on whether the project objectives were achieved. A technique for writing an objective is to make sure it is specific, measurable, attainable/achievable, realistic, and time bound (SMART).

One-Piece Flow: Making and moving only one piece or part at a time. See Continuous Flow.

One-Sided Alternative: The value of a parameter that has an upper bound or a lower bound, but not both.

Operator Cycle Time: The time it takes an operator to go through all of his or her work elements before repeating them.

Order Interval: Represents the frequency (days) that a part is ordered.

Ordinal: Ordered categories (ranking) with no information about distance between each category, that is, rank ordering of several measurements of an output parameter.

Ordinate: The vertical axis of a graph.

Overall Equipment Effectiveness (OEE): A total productive maintenance (TPM) measure of how effectively equipment is being used. OEE = availability rate × performance rate × quality rate.

Overproduction: The process of producing more, sooner, or faster than is required by the next process or customer.

P Charts: Charts used to plot percent of defectives in a sample.

Pacemaker: The only point in the production process that is scheduled, and therefore, dictates the pace of production for a whole system of processes.

Pacesetter: The point in the process that limits the output of the total process.

Pair Programming: An Agile software development technique in which two programmers work together at one workstation. One types in code while the other reviews each line of code as it is typed in. The person typing is called the driver. The person reviewing the code is called the observer (or navigator). The two programmers switch roles frequently.

Parameter: A constant defining a particular property of the density function of a variable.

Pareto Diagram: A chart that ranks, or places in order, common occurrences.

Perturbation: A nonrandom disturbance.

Pitch: The amount of time required by a production area to make one container of product. Takt time × pack-out qty = pitch.

Plan, Do, Check, Act (PDCA): An improvement cycle introduced to the Japanese in the 1950s by W. Edwards Deming. Based upon proposing, then implementing, an improvement, then measuring the results and acting accordingly.

Plan for Every Part (PFEP): A comprehensive plan for each part consumed within a production process. This would take the form of a spreadsheet or simple table and contain such data as pack-out quantity, location of use and storage, order frequency, and so on. This provides one accurate source of information relating to parts.

Planning Poker: Also called Scrum poker, is a consensus-based technique for estimating, mostly used to estimate effort or relative size of tasks in software development.

Poka-Yoke: Mistake-proof device or procedure designed to prevent a defect from occurring throughout the system or process. Error-proofing is a manufacturing technique of preventing errors by designing the manufacturing process, equipment, and tools so that an operation literally cannot be performed incorrectly. Poka-yoke is the Japanese phrase for "do it right the first time."

Population: A group of similar items from which a sample is drawn. Often referred to as the universe.

Power of an Experiment: The probability of rejecting the null hypothesis when it is false and accepting the alternative hypothesis when it is true.

Prevention: The practice of eliminating unwanted variations of priori (before the fact), for example, predicting a future condition from a control chart and when applying corrective action before the predicted event transpires.

Primary Control Variables: The major independent variables in the experiment.

Probability: The chance of something happening in percent or number of occurrences over a large number of trials.

Probability of an Event: The number of successful events divided by the total numbers of trials.

Problem: A deviation from a specified standard.

Problem Solving: A process of solving problems, the isolation and control of those conditions that generate or facilitate the creation of undesirable symptoms.

Process: A particular method of doing something, generally involving a number of steps or operations.

Process Average: The central tendency of a given process characteristic across a given amount of time or a specific point in time.

Process Control: See Statistical Process Control.

Process Control Chart: Any of a number of various types of graphs upon which data are plotted against specific control limits.

Process Owner: They have the responsibility for process performance and resources. They provide support, resources, and functional expertise to Six Sigma projects. They are accountable for implementing developed Six Sigma solutions into their process.

Process Spread: The range of values that a given process characteristic displays; this particular term most often applies to the range but may also encompass the variance. The spread may be based on a set of data collected at a specific point in time or may reflect the variability across a given amount of time.

Producers Risk: Probability of rejecting a lot when, in fact, the lot should have been accepted. See Alpha Risk.

Product Backlog: A prioritized list of everything that needs to be done to complete a project.

Product Owner: A term commonly used in Scrum (Agile) denoting one of the project's key stakeholders. The product owner's responsibility includes envisioning what should be built or created and conveying that to the team.

Production Kanban: A signal that specifies the type and quantity of product that an upstream process must produce.

Program: A program is the umbrella structure established to manage a series of related projects. The program does not produce any project deliverables. The project teams produce them all. The purpose of the program is to provide overall direction and guidance, to make sure the related projects are communicating effectively, to provide a central point of contact and focus for the client and the project teams, and to determine how individual projects should be defined to ensure that all the work gets completed successfully.

Program Manager: A program manager is the person with the authority to manage a program. (Note that this is a role. The program manager may also be responsible for one or more of the projects within the program.) The program manager leads the overall planning and management of the program. All project managers within the program report to the program manager.

Project: A project is a temporary structure to organize and manage work and ultimately to build a specific defined deliverable or set of deliverables. By definition, all projects are unique, which is one reason it is difficult to compare different projects to one another. A problem usually calling for planned action.

Project Definition (Charter): Before starting a project, it is important to know the overall objectives of the project as well as the scope, deliverables, risks, assumptions, project organization chart, etc. The project definition (or charter) is the document that holds this relevant information. The project manager is responsible for creating the project definition. The document should be approved by the sponsor to signify that the project manager and the sponsor agree on these important aspects of the project.

Project Manager: The project manager is the person with the authority to manage a project. The project manager is 100% responsible for the processes used to manage the project. He or she also has people management responsibilities for team members although this is shared with the team member's functional manager. The processes used to manage the project include defining the work, building the work plan and budget, managing the work plan and budget, scope management, issues management, risk management, etc.

Project Phase: A phase is a major logical grouping of work on a project. It also represents the completion of a major deliverable or set of related deliverables. On an IT development project, logical phases might be planning, analysis, design, constructing (including testing), and implementation.

Project Team: The project team consists of the full-time and part-time resources assigned to work on the deliverables of the project. They are responsible for understanding the work to be completed; completing assigned work within the budget, timeline, and quality expectations; informing the project manager of issues, scope changes, and risk and quality concerns; and proactively communicating status and managing expectations.

Pull: Material flow triggered by actual customer need rather than a scheduled production forecast. Downstream processes signal to upstream processes exactly what is required and in what quantity.

Push: The production of goods regardless of demands or downstream need, usually in large batches to ensure efficiency.

Quality Function Deployment (QFD): A systematic process used to integrate customer requirements into every aspect of the design and delivery of products and services.

R Charts: Plot of the difference between the highest and lowest in a sample range control chart.

Random: Selecting a sample so each item in the population has an equal chance of being selected, lack of predictability.

Random Cause: A source of variation that is random; a change in the source ("trivial many"), for example, a correlation does not exist, any individual source of variation results in a small amount of variation in the response, cannot be economically eliminated from a process, an inherent natural source of variations.

Random Effects Model: Experimental treatments are a random sample from a larger population of treatments. Conclusion can be extended to the population. Interferences are not restricted to the experimental levels.

Random Sample: One or more samples randomly selected from the universe (population).

Random Variable: A variable that can assume any value of a set of possible values.

Random Variations: Variations in data that result from causes that cannot be pinpointed or controlled.

Randomness: A condition in which any individual event in a set of events has the same mathematical probability of occurrence as all other events within the specified set, that is, individual events are not predictable even though they may collectively belong to definable distribution.

Range: The difference between the highest and lowest in a set of values or "subgroup."

Ranks: Values assigned to items in a sample to determine their relative occurrence in a population.

Ratio: A numeric scale that has an absolute zero point and equal units of measure through, that is, measurements of an output parameter, for example, amps.

Regression Analysis: Includes any techniques for modeling and analyzing several variables. Linear regression was the first type of regression analysis to be studied rigorously and to be used extensively in practical applications.

Reject Region: The region of values in which the alternative hypothesis is accepted.

Repeatability (of a Measurement): The extent to which repeated measurements of a particular object with a particular instrument produce the same value.

Replication: Observations made under identical test conditions.

Representative Sample: A sample that accurately reflects a specific condition or set of conditions within the universe.

Reproducibility (of a Measurement): The extent to which repeated measurements of a particular object with a particular individual produce the same value.

Requirements: Requirements are descriptions of how a product or service should act, appear, or perform. Requirements generally refer to the features and functions of the deliverables that are building on the project. Requirements are considered to be a part of project scope. High-level scope is defined in the project definition (charter). The requirements form the detailed scope. After the requirements are approved, they can be changed through the scope change management process.

Research: Critical and exhaustive investigation or experimentation having for its aim the revision of an accepted conclusion in the light of newly discovered facts.

Residual Error: See Experimental Error.

Response Time: The time in which an order needs to be satisfied.

Retrospective: A team meeting that happens at the end of every development iteration to review lessons learned and to discuss how the team can be more efficient in the future. It is based on the principles of applying the learning from the previous sprint to the upcoming sprint.

Rework: Activity required to correct defects produced by a process.

Risk: There may be potential external events that will have a negative impact on the project if they occur. Risk refers to the combination of the probability the event will occur and the impact on the project if the event occurs. If the combination of the probability of the occurrence and the impact to the project is too high, the potential event should be identified as a risk and a proactive plan should be put in place to manage the risk.

Robust: The conditions or state in which a response parameter exhibits hermetic to external cause of a nonrandom nature, that is, impervious to perturbing influence.

Safety Stock: Inventory held to compensate for variation in demand, quality, and downtime.

Sample: One or more observations drawn from a larger collection of observations or universe (population).

Scatter Diagrams: Charts that allow for the study of correlation, for example, the relationship between two variables.

Scope: Scope is the way the boundaries of the project are described. It defines what the project will deliver and what it will not deliver. High-level scope is set in the project definition (charter) and includes all of the deliverables and the boundaries of the project. The detailed scope is identified through the business requirements. Any changes to the project deliverables, boundaries, or requirements would require approval through scope change management.

Scope Change Management: The purpose of scope change management is to manage change that occurs to previously approved scope statements and requirements. Scope is defined and approved in the scope section of the project definition (charter) and the more detailed business requirements. If the scope or the business requirements change during the project (which usually means that the client wants additional items), the estimates for cost, effort, and duration may no longer be valid. If the sponsor agrees to include the new work in the project scope, the project manager has the right to expect that the current budget and deadline will be modified (usually increased) to reflect the additional work. This new estimated cost, effort, and duration now become the approved target. Sometimes the project manager thinks that scope management means having to tell the client "no." That makes the project manager nervous and uncomfortable. However, the good news is that managing scope is all about getting the sponsor to make the decisions that will result in changes to project scope.

Scrum: Scrum is a popular framework for putting Agile methods into practice.

Scrum Board: Also called Kanban, displaying a sticky note for each task in progress. These are aligned in separate columns based on their status: to-do, doing, or done.

Scrum Master: A person who helps teams manage themselves, ensuring they have all the resources and information they need.

Sensei: Japanese word for "teacher" and denotes mastery within their field of knowledge. A sensei should be a wise and easily understood mentor that guides thinking with his subjects rather than dictating the point so as to promote learning.

Setup Time: The amount of time required to changeover a process after producing the last part of one product to the first part of the next product.

Short-Term Variation: The amount of variation observed in a characteristic that has not had the opportunity to experience all the sources of variation from the inputs acting on it.

Signal Kanban: A signal that triggers an upstream process to produce when a minimum quantity is reached at the downstream process.

Single Minute Exchange of Die (SMED): A technique to reduce setup or changeover times to eliminate the need to build in batches.

Spaghetti Chart: A visual chart showing the path taken by a product or person during a process to highlight excessive motion.

Special Cause: See Assignable Cause.

Special Cause Variation: Nonrandom causes of variation, sometimes outside the project manager's control.

Specification Limits: The boundaries of acceptable performance.

Spike: A short, time-boxed piece of research, usually technical, on a single story that is intended to provide just enough information that the team can estimate the size of the story.

Sponsor (Executive Sponsor and Project Sponsor): The sponsor is the person who has ultimate authority over the project. The executive sponsor provides project funding, resolves issues and scope changes, approves major deliverables, and provides high-level direction. He or she also champions the project within the organization. Depending on the project and the organizational level of the executive sponsor, he or she may delegate day-to-day tactical management to a project sponsor. If assigned, the project sponsor represents the executive sponsor on a day-to-day basis and makes most of the decisions requiring sponsor approval. If the decision is large enough, the project sponsor will take it to the executive sponsor.

Sprint/Iteration: A work period of a fixed length, usually four to six weeks.

Sprint Planning: This is a pre-sprint planning meeting attended by the core Agile team. During the meeting, the product owner describes the highest priority features to the team as described on the product backlog. The team then agrees on the number of features that can be accomplished in the sprint and plans out the tasks required to achieve delivery of those features. The planning group works the features into user stories and assigns acceptance criteria to each story.

Sprint Review: Each sprint is followed by a sprint review. During this review, the software developed in the previous sprint is reviewed and, if necessary, new backlog items are added.

Stable Process: A process that is free of assignable causes, for example, in statistical control.

Stakeholder: Specific people or groups who have a stake in the outcome of the project are stakeholders. Normally, stakeholders are from within the company and may include internal clients, management, employees, administrators, etc. A project can also have external stakeholders, including suppliers, investors, community groups, and government organizations.

Standard Deviation: One of the most common measures of variability in a data set or in a population. It is the square root of the variance.

Standardized Work: A defined work method that describes the proper workstation and tools, work required, quality, standard inventory knacks, and sequence of operations.

Statistical Control: A quantitative condition that describes a process that is free of assignable/special causes of variation, for example, variation in the central tendency and variance. Such a condition is most often evidence on a control chart, that is, a control chart that displays an absence of nonrandom variation.

Statistical Process Control (SPC): The use of basic graphical and statistical methods for measuring, analyzing, and controlling the variation of a process for the purpose of continuously improving the process.

Steering Committee: A steering committee is usually a group of high-level stakeholders who are responsible for providing guidance on overall strategic direction. They don't take the place of a sponsor but help spread the strategic input and buy-in to a larger portion of the organization. The steering committee is especially valuable if the project has an impact in multiple organizations as it allows input from those organizations into decisions that affect them.

Story/Stories: See User Stories.

Story Points: Unit of estimation measuring complexity and size.

Subgroup: A logical grouping of objects or events that display only random event-to-event variations, for example, the objects or events are grouped to create homogenous groups free of assignable or special causes. By virtue of the minimum within group variability, any change in the central tendency or variance of the universe will be reflected in the subgroup-to-subgroup variability.

Supermarket: A strategically controlled store of parts used by downstream processes.

Supplier: A vendor or entity responsible for providing an input to a process in the form of resources or information.

Symptom: That which serves as evidence of something not seen.

System: That which is connected according to a scheme.

Systematic Variables: A pattern that displays predictable tendencies.

Takt Time: Rate of demand from customer. It is the available operating time for the requirement.

Task: A user story that can be broken down into one or more tasks. Tasks are estimated daily in hours (or story points) remaining by the developer working on them.

Task Board/Storyboard: A wall chart with cards and sticky notes that represents all the work for a given sprint. The notes are moved across the board to show progress.

Team: The team is responsible for delivering the product. A team is typically made up of five to nine people with cross-functional skills who do the actual work (analyze, design, develop, test, technical communication, document, etc.). It is recommended that the team be self-organizing and self-led, but often work with some form of project or team management.

Test of Significance: A procedure to determine whether a quantity subjected to random variation differs from postulated value by an amount greater than that due to random variation alone.

Test-Driven Development: Test-driven development (TDD) is a software development process that relies on the repetition of a very short development cycle First, the developer writes a failing automated test case that defines a desired improvement or new function, then produces code to pass that test and finally refactors the new code to acceptable standards.

Theory: A plausible or scientifically acceptable general principle offered to explain phenomena.

Theory of Constraints: Theory of constraints describes the methods used to maximize operating income when an organization is faced with bottleneck operations. This theory also deals with how to handle the unknown.

Time Box/Boxing: A maximum period of time allotted to produce something of value to the customer.

Total Cycle Time (TCT): The time taken from work order release into value stream until completion/movement of the product into shipping/finished goods.

Total Productive Maintenance: A means of maximizing production system efficiency by analyzing and eliminating downtime through upfront maintenance of equipment.

Toyota Production System: The production system developed and used by the Toyota Motor Company that focuses on the elimination of waste throughout the value stream.

Trend: A gradual, systematic change over time or some other variable.

Two-Sided Alternative: The value of a parameter that designates an upper and lower bound.

Type I Error: See Alpha Risk.

Type II Error: See Beta Risk.

Unnatural Pattern: Any pattern in which a significant number of the measurements do not group themselves around a center line; when the pattern is unnatural it means that outside disturbances are present and are affecting the process.

Upper Control Limit: A horizontal line on a control chart (usually dotted) that represents the upper limits of process capability.

User Persona: Personas are a description of the typical users of a given software. A persona description should include skills, background, and goals.

User Story: A user story is a very high-level definition of a requirement, containing just enough information so that the developers can produce a reasonable estimate of the effort to implement it. A user story is one or more sentences that capture what the users want to achieve. A user story is also a placeholder for conversation between the users and the team. The user stories should be written by or for the customers for a software project and are the main instrument to influence the development of the software. User stories could also be written by developers to express nonfunctional requirements (security, performance, quality, etc.). An easier way of thinking of about user stories is that they are narratives defining features, functions, and other work to be delivered, explaining who needs the task and why.

Value: This term refers to a product or service capability that is provided to a customer at the right time and at an appropriate price.

Value-Added Activity: Any activity that changes the product in terms of fit, form, or function toward something that a customer is willing to pay for.

Value-Added Time: The time expanded in value-added activity to produce a unit. Time for those work elements that transform the product in a way that the customer is willing to pay for.

Value Stream: All activities, both value-added and non–value added, that are required to bring a product, group, or service from the point of order to the hands of a customer and a design from concept to launch to production to delivery.

Value Stream Map: A visual representation of a process showing flow of information and material through all steps from the supplier to the customer.

Variable: A characteristic that may take on different values.

Variables Data: Numerical measurement made at the interval or ratio level; quantitative data, for example, ohms, voltage, diameter, or subdivision of the measure scale are conceptually meaningful, for example, 1.6478 volts.

Variation: Any quantifiable difference between individual measurements; such differences can be classified as being due to common causes (random) or special causes (assignable).

Variation Research: Procedures, techniques, and methods used to isolate one type of variation from another (for example, separating product variation from test variation).

Velocity: A relative number that describes how much work the team can get done over a period of time.

Visualization: The design of a workplace such that problems and issues can be identified without timely and in-depth investigation. Truly visual workplaces should be capable of assessment in less than three seconds.

VOB (Voice of Business): The voice of the business is derived from financial information and data. Voice of the business represents the needs of the business and the key stakeholders of the business. It is usually items such as profitability, revenue, growth, market share, etc.

VOC (Voice of Customer): Voice of the customer represents the expressed and non-expressed needs, wants, and desires of the recipient of a process output, a product, or a service. It is usually expressed as specifications, requirements, or expectations.

VOE (Voice of Employee): Voice of the employee represents the expressed and non-expressed needs, wants, and desires of what the employee needs to be successful.

VOP (Voice of Process): Voice of the process represents the performance and capability of a process to achieve both business and customer needs.

Waste (Muda): Includes anything that does not add value to a final product or service, an activity that consumes valuable resources without creating customer value.

Waterfall method: A traditional method of organizing projects in steps. Typically Step One needs to be complete before moving on to Step Two.

WIP (Work in Process): These are items—material or information—that are between machines, processes, or activities waiting to be processed, any inventory between raw materials and finished goods.

Withdrawal Kanban: A signal that specifies the type and quantity of product that the downstream process may withdraw.

Work Cells: An arrangement of people, machines, materials, and methods such that processing steps are adjacent and in sequential order; thus parts can be processed one at a time.

Work Plan (Schedule): The project work plan describes how the project will be completed. It describes the activities required, the sequence of the work, who is assigned to the work, an estimate of how much effort is required, when the work is due, and other information of interest to the project manager. The work plan allows the project manager to identify the work required to complete the project and also allows the project manager to monitor the work to determine whether the project is on schedule.

X: Input.

X&R Charts: A control chart that is a representation of process capability over time, displays variability in the process average and range across time.

XP: A software development methodology that is intended to improve software quality and responsiveness to changing requirements. As a type of Agile software development, it advocates frequent "releases" in short development cycles (time boxing), which is intended to improve productivity and introduce checkpoints at which new customer requirements can be adopted.

Y: Output.

Appendix C: Lean Six Sigma Competency Models

Lean Six Sigma Yellow Belt

Basic Competency Model

Competency Performance Criteria

High-level understanding of the following:

- Basic DMAIC (Define, Measure, Analyze, Improve, Control) concept
- PDCA (plan–do–check–act) model
- How Lean and Six Sigma work together

Ability to Explain the General Roles and Responsibilities of Lean Six Sigma Human Resources to Include

- Master black belt
- Black belt
- Green belt
- Yellow belt
- White belt
- Champion
- Sponsor
- Process owner

Ability to Identify the 7 Tools of Quality and Their Overall Purpose

- Fishbone
- Check sheet
- Flow chart
- Histogram
- Pareto chart
- Scatter diagram
- Control chart

Exposure/Understanding of Basic Project Management

- Project charter
- Process mapping
- Opening and closing a project
- Basic project management tools

Understanding of the Importance of the Following as It Relates to Lean Six Sigma:

- VOC, VOB, VOE, and VOP (voices of customer, business, employee, and process)
- SIPOC model (supply–input–process–output–customer)
- CTQ (critical to quality)
- Benchmarking

Ability to Explain Why Lean Six Sigma Is Important for Process Improvement and How It Relates to Other Process Improvement Programs

SSD Global supports the concept that all process improvement programs are rooted in Total Quality Management (TQM) concepts and that process improvement begins with a firm understanding of Project Management basics.

Lean Six Sigma Green Belt

Basic Competency Model

Competency Performance Criteria

Ability to Define Lean Six Sigma

- Philosophy of Lean Six Sigma
- Overview of DMAIC (Define, Measure, Analyze, Improve, Control)
- Understand how Lean and Six Sigma work together

Ability to Explain the Roles and Responsibilities of Lean Six Sigma Participants

- Master black belt
- Black belt
- Green belt
- Yellow belt
- White belt
- Champion
- Executive
- Coach
- Facilitator
- Team member
- Sponsor
- Process owner

Ability to Use the 7 Tools of Quality

- Fishbone
- Check sheet
- Flow chart
- Histogram
- Pareto chart
- Scatter diagram
- Control chart

Exposure to Basic Project Management

■ Project charter
■ Process mapping
■ Opening and closing a project
■ Basic project management tools

Describe the Impact that Lean Six Sigma Has on Business Operations

■ Methodologies for improvement
■ Theories of VOC, VOB, VOE, and VOP

Ability to Identify and Explain Areas of Waste:

■ Excess inventory
■ Space
■ Test inspection
■ Rework
■ Transportation
■ Storage
■ Reducing cycle time to improve throughput
■ Skills

Ability to Explain Why Lean Six Sigma Is Important for Process Improvement

SSD Global supports the concept that all process improvement programs are rooted in Total Quality Management (TQM) concepts and that process improvement first begins with a firm understanding of Project Management basics as outlined in the Project Management Body of Knowledge (PMBOK®). Lean Six Sigma Green Belts should begin studying these areas. SSD Global suggests that Lean Six Sigma practitioners consider joining the Project Management Institute and/or the American Society of Quality.

Lean Six Sigma Black Belt

Basic Competency Model

Criteria for Testing and Practical Application

Ability to Lead a DMAIC Project

- Complete understanding of the Define–Measure–Analyze–Improve–Control process
- Understand leadership responsibilities in deploying a Lean Six Sigma project
- Understand change management models
- Be able to communicate ideas

Ability to Describe and Identify Organizational Roadblocks and Overcome Barriers

- Lack of resources
- Management support
- Recovery techniques
- Change management techniques

Using tools and theories such as

- Constraint management
- Team formation theory
- Team member selection
- Team launch
- Motivational management

Understand Benchmarking, Performance, and Financial Measures:

- Best practice
- Competitive
- Collaborative
- Score cards
- COQ/COPQ
- ROI
- NPV

Use and Understand the Following Lean Six Sigma Tools:

- Check sheets
- Control charts (line and run charts) and be able to analyze typical control chart patterns
- Critical path
- Fishbone
- Flowcharting
- FMEA
- Gantt chart
- Histogram
- Pareto chart
- PERT chart
- Scatter diagrams
- Spaghetti diagrams
- Swim lane charts
- SWOT analysis
- TIM WOODS or the Eight Areas of Waste
- Value stream mapping (basic)

Define and Distinguish between Various Types of Benchmarking, Including Best Practices, Competitive, and Collaborative.

Define Various Business Performance Measures, Including Balanced Scorecard, Key Performance Indicators (KPI), and the Financial Impact of Customer Loyalty.

Define Financial Measures such as Revenue Growth, Market Share, Margin, Cost of Quality (COQ), Net Present Value (NPV), Return on Investment (ROI), and Cost-Benefit Analysis

More detailed information is contained in the Lean Six Sigma Body of Knowledge (SSD Global Version 3.1 available online at http://www.SSDGlobal.net).

SSD Global supports the concept that all process improvement programs are rooted in Total Quality Management (TQM) concepts and that process improvement first begins with a firm understanding of Project Management basics as outlined in the Project Management Body of Knowledge (PMBOK). Lean Six Sigma Black Belts and Master Black Belts should be well versed in these areas. SSD Global suggests that Lean Six Sigma practitioners consider joining the Project Management Institute and/or the American Society of Quality.

SSD Global further supports that the newer and leaner Lean Six Sigma, which is based on Six Sigma with a heavy emphasis in Lean Manufacturing/ Lean Thinking, has evolved to include other established bodies of knowledge. In addition to basic TQM and the PMBOK, successful Lean Six Sigma Black Belts and Master Black Belts should review, study, and monitor these additional bodies of knowledge:

■ Business Analysis Body of Knowledge (BABOK®)
■ Business Process Reengineering (BPR)
■ Change Management
■ Leadership Development
■ Measurement Systems Analysis
■ Statistics
■ Business Finance

Organizational Development

Lean Six Sigma Master Black Belt

Basic Competency Model

Professional competency models are established to provide guidelines in determining expertise and knowledge in a particular area or subject. The following criteria may be used for interview questions, testing, and practical application exercises.

Ability to Identify and Lead a DMAIC Project

■ Ability to teach and facilitate the Define–Measure–Analyze–Improve–Control process
■ Demonstrate leadership in deploying a Lean Six Sigma project
■ Deploy and monitor change management models
■ Superior verbal and written presentation skills

Ability to Creatively Deal with Roadblocks and Overcome Barriers Related to

■ Lack of resources
■ Management support
■ Recovery techniques
■ Change management techniques

Teaching and Mentoring Knowledge of Tools and Theories to Include

- Constraint management
- Team formation theory
- Team member selection
- Team launch
- Motivational management

Prepare, Explain, and Evaluate Factors Related to Benchmarking, Performance, and Financial Measures:

- Best practice
- Competitive
- Collaborative
- Score cards
- Cost of quality/cost of poor quality (COQ/COPQ)
- Return on investment (ROI)
- Net present value (NPV)

Use, Evaluate, and Explain:

- Check sheets
- Control charts (line and run charts) and be able to analyze typical control chart patterns
- Critical path
- Fishbone
- Flowcharting
- FMEA
- Gantt chart
- Histogram
- Pareto chart
- PERT chart
- Scatter diagrams
- Spaghetti diagrams
- Swim lane charts
- SWOT analysis
- TIM WOODS or the Eight Areas of Waste
- Value stream mapping (Basic)

Develop, Delivery, Evaluate Training Plans

■ Design training plans
■ Understand various training approaches
■ Build curriculum
■ Demonstrate success
■ Be able to coach and mentor

Additional Design Criteria

■ Business performance measures such as
  – Balanced scorecard
  – Key performance indicators (KPI)
  – Financial measures
    • Revenue growth
    • Market share
    • Margin
    • Cost of quality (COQ)/cost of poor quality (COPQ)
    • Net present value (NPV)
    • Return on investment (ROI)
    • Cost-benefit analysis

***

SSD Global supports the concept that all process improvement programs are rooted in Total Quality Management (TQM) concepts and that process improvement first begins with a firm understanding of Project Management basics as outlined in the Project Management Body of Knowledge (PMBOK). The International Lean Six Sigma Master Black Belt should be well versed in these areas. SSD Global also suggests that International Lean Six Sigma Master Black Belt familiarize themselves with ISO 13053, ISO 12500, and PRINCE2®.

Index

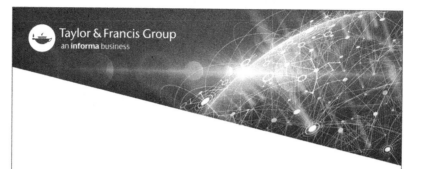

Printed in the United States
by Baker & Taylor Publisher Services